KEY NOTES ON
FOOD SCIENCE AND TECHNOLOGY

For Ready Reference to the

STUDENTS, TEACHERS, RESEARCHERS & ASPIRANTS OF COMPETITIVE EXAMINATIONS

THE EDITORS

Dr. U.D. Chavan obtained his M.Sc. (Agri. in Biochemistry) degree from Mahatma Phule Krishi Vidyapeeth, Rahuri. He received his Ph.D. degree in Food Science from Memorial University of Newfoundland St. John's Canada in 1999. He has done International Training on "Global Nutrition 2002" at Uppsala University Uppasala, Sweden in 2002. Dr. Chavan worked as Senior Research Assistant in the Department of Biochemistry & Food Science and Technology at MPKV Rahuri from 1988 to 2000. During his Ph.D., he worked as Technician/Research Associate at Atlantic Cool Climate Crop Research Center and Agriculture and Agri-Food Canada. He received D.Sc. degree in 2006 from USA.

Dr. Chavan is presently working as a Senior Cereal Food Technologist in the Department of Food Science & Technology at Mahatma Phule Krishi Vidyapeeth, Rahuri.

Dr. J.V. Patil obtained his M.Sc. (Agri.) from, MPKV, Rahuri. He completed his course work for Ph.D. at CCSHAU, Hisar and research at MPKV, Rahuri in 1992. He rendered his research and teaching services at MPKV Rahuri as Geneticist, Associate Professor, Plant Breeder and Professor of Genetics & Plant Breeding and Head, Genetics and Plant Breeding Department, MPKV, Rahuri. He also delivered many administrative responsibilities in the University. Dr. Patil joined as the Director, Directorate of Sorghum Research, Hyderabad in August 2010.

THE CONTRIBUTOR

Dr. U.S. Dalvi is an Assistant Professor in the Department of Biochemistry at Mahatma Phule Krishi Vidyapeeth, Rahuri.

KEY NOTES ON
FOOD SCIENCE AND TECHNOLOGY

For Ready Reference to the

**STUDENTS, TEACHERS, RESEARCHERS & ASPIRANTS OF COMPETITIVE
EXAMINATIONS**

Editors

U.D. CHAVAN
&
J.V. PATIL

Contributor

U.S. DALVI

2015

Daya Publishing House®
A Division of
Astral International (P) Ltd
New Delhi 110 002

© 2015 PUBLISHER
ISBN: 9789351307044 (International Edition)

Published by   : **Daya Publishing House®**
                        A Division of
                        Astral International Pvt. Ltd.
                        – ISO 9001:2008 Certified Company –
                        4760-61/23, Ansari Road, Darya Ganj
                        New Delhi-110 002
                        Ph. 011-43549197, 23278134
                        E-mail: info@astralint.com
                        Website: www.astralint.com

Laser Typesetting   : **Twinkle Graphics, Delhi**

Printed at   : **Thomson Press India Limited**

PRINTED IN INDIA

PREFACE

India is an agricultural country. The Indian economy is basically agarian. Inspite of economic and industrialization, agriculture is the backbone of the Indian economy. As Mahatma Gandhi said "India's lives in villages and agriculture is the soul of Indian economy". Agriculture is a vast subject and encompasses at least 20 major and minor subjects in it. New developments have lead to entirely a new face of agriculture. Study of agriculture has always been intrigued with a mosaic of interwove concepts, subjects, facts and figures. There are number of books and large literature on Food Science and Technology but the Key Notes type of book have not been compiled in a readable manner.

The present book *"Key Notes on Food Science and Technology"* has been designed to fulfill this long felt need of students, teachers, researchers and aspirants of competitive examinations. It is designed in such a way that give rapid, easy access to the core materials in a short format which facilitates easily learning and rapid revision. The book carries fundamentals of Food Science and Technology. There are seven chapters elaborating Discoveries, Abbreviations, Terminology, Distinguish/Comparison, Short explanations, Food Biotechnology, Human Nutrition and Dietetics as well as references also included. The most recent information is provided along with a detailed list of references for further reading.

Hope this book would be highly useful for graduate and post-graduate students of agriculture, teachers and researchers. This book will also useful for the aspirants of various competitive examinations such as Agricultural Research Service (ARS), ICAR- National Eligibility Test (NET), State Eligibility Test (SET), Junior Research Fellowship (JRF), Senior Research Fellowship (SRF), Civil Services, Allied Agricultural Examinations and Extension Workers for reference and easy answers of many complicated questions. Thus it is expected that this book will adequately meet the need of wider circle of students and readers for preparing their professional career.

We acknowledge the references that are used in this manuscript. Authors are also thankful to all scientists and friends who have helped directly or indirectly while preparing this manuscript. The editors of grateful to all the contributors for their cooperation, support and timely submission of their manuscripts for bringing out this publication. We would have like to acknowledge the patience and support of our families whilst we have spent many hours with drafts of manuscripts rather than with them. Lastly, our sincere thanks to publisher Astral International Pvt. Ltd., New Delhi who provides an opportunity to publish this book.

To all readers we extend an invitation to report that no doubts have escaped our attention and to offer suggestion for improvements that can be incorporated in future editions.

U.D.Chavan and J.V. Patil

Editors

CONTENTS

1

DISCOVERIES

Scientist	Year	Discovery
Greek Physician Hippocrates	300 BC	Father of medicine had many ideas about nutrition and diets
Drummond	1400 BC	Noted night blindness
Carl Wilhelm & Scheele	1742-86	Isolated & studied properties of lactic acid, Discovered chlorine, glycerol and oxygen 3 years before priestly but unpublished
Antoine Laurent & Lavoisier	1743-94	Principles of Modern Chemistry
Marggraf	1747	Milk sugar
Needham	1749	First time he explained the cause of spoilage of stored food
James Lind sailors	1753	Scurvy can be cured by fresh fruits and vegetables among
Spallanzani	1765	Microorganisms are responsible for spoilage of fruits and vegetables during storage
Nicolas	1767-1845	Principles of Agricultural and Food Chemistry
Lavoisier	1770-94	Nature of respiration, oxidation of organic compounds into CO_2, H_2O and heat (energy)
Josephlouis, Gay vegetable -Lussac & Louis-Jacques, Thenard	1778-1850	Percentage of carbon, hydrogen and nitrogen in dry substances
Jons Jacob Berzelius & Thomas Thomson	1779-1848 1773-1852	Beginnings of organic formulas
William Beaumont	1785-1853	Gastric digestion
Michel Eugene & Chevreul	1786-1889	Listed O, Cl, I, N, S, P, C, Si, H, Al, Mg, Ca, Na, K, Mn, Fe, exist in organic substances
Fr.de. Fourcroy	1789	Identified proteins and fats in plants and animal tissues
Arthur Hill Hassall	1800	Differentiated pure and adulterated foodstuffs
Jean Baptiste Dumas	1800-84	Only protein, carbohydrate and fat are inadequate for support of life
Beaccornot	1802	First hydrolyzed protein with acid

Scientist	Year	Discovery
Justus Von Liebig	1803-73	Vinegar fermentation, First book on Food Chemistry (1847), Research on the chemistry of Food. He also discovered nitrogenous and non-nitrogenous compounds in food
M. Nicholas	1804	First to report the successful preservation of food in glass containers
Prout	1806	Glucose and malt sugar and protein analysis
Sir Humphry Davy	1807-08	Isolated K, Na, Ba, Sr, Ca and Mg
M. Nicholas	1809	Awarded the prize; Book on "The art of Preserving Animal and Vegetable Substances for Many Years"
Wollaston	1810	Discovered cystine (amino acid)
Courtois	1811	Presence of iodine in burnt sponge (for the treatment of goiter)
Chevereul	1814	Fats are composed of fatty acids and glycerol
Fastier	1824	"Hole-and-cap" cans development; patent granted in 1839
Angilbert	1833	Full aperture containers
Prout	1834	Put forward the theory that there are three nutrients in food: CHO, Fats and Proteins
Mulder	1838	The word protein was coined
Leibig oxidized	1842	Not organic compounds but carbohydrates, fats and proteins
Winslow and Raymond Chevalien Appert	1843	Canned foods could be processed by steam and water under pressure
-	1852	Development of pressure cookers
Takaki	1857	Observed addition of meat, vegetables, condensed milk and barley in the diet of sailors prevent beriberi disease
-	1857	Fruit and vegetable processing first started in an organized manner
In Germany	1860	Established first Agricultural experiment station, Hanneberg, W. - Director & Stohmann, F. - Chemist
Papin	1861	First time cooking of foods by means of pressure
Isaac Newton	1862	Established United States Department of Agriculture (USDA), First Commissioner
Harvey Washington Wiley	1863	First Chief Chemist of the USDA
Louis Pasteur	1864	Role of microorganisms in food spoilage
Peltier and Paillard	1868	Used varnish for internal coating of cans
Shriver	1874	Autoclave provided with inlet for steam from external source
Howe	1876	Soldering of cans
Parry and Cobley	1882	Used sodium, potassium or calcium silicate and a serum made of proteinous materials
Baumann	1896	Presence of iodine in thyroid gland

Scientist	Year	Discovery
Government Policy	1906	Pure Food and Drug Act (PFDA)
Hopkins, Osborne and Mendel	1906 1912	Showed zein a maize protein did not support for growth, deficient in lysine and tryptophan
Funk	1912	Coined the term Vitamin for the growth factors, Vitamin theory
McCollum & Davis	1913	Vitamin "A"
McCollum & Coworkers	1913	Water soluble Vitamin "B"
Cold berger	1915	Found that the addition of milk and eggs to poor maize diet to prevent occurrence of pellagra
-	1927	Canning of fruits and vegetables started for export
Filby, F. A.	1934	A history of food adulteration and analysis
Cicely Williams	1935	Demonstrated Kwashiorkor disease in children due to protein deficiency and can be cured by feeding milk
Farrow and Green	1941	Classified lacquers into five groups
Government of India	1935	First fruit and vegetable processing factory was established at Bombay
Government of Uttar Pradesh	1949	Fruit Preservation and Canning Institute was established at Lucknow
Government of India	1950	Established Central Food Technological Research Institute, Mysore
Government of India	1955	Passed Fruit Products Order
Government of India	1973	For licensing, a Food and Nutrition Board was established
Darby	1977	Noted several deficiencies in human health in ancient Egypt
Van Leeuwenhock	-	Developed microscope
Harvey	-	Blood circulation in body
Rutherford	-	Discovered nitrogen
Priestly	-	Discovered oxygen
Black	-	Carbondioxide
Lavoisier	-	Showed that respiration is the life process
Von Liebig	-	Founder of the sciences of Agricultural chemistry and with Pasteur of Biochemistry
Lunin	-	Vitamins in milk
Eijkman	-	Vitamins in rice
Szent-Gyorgyi, King & Waugh	-	Isolation and identification of Vitamin "C"
Lavoisier	-	Established the law of conservation of mass. Father of modern chemistry and Nutrition
Scottish Physician	-	Lime and lemons are good source to prevent scurvy disease

2

ABBREVIATIONS

Abbreviation	Full Form
1-MCP	1-methyl cyclopropene
AACC	American Association of Cereal Chemists
AAO	Ascorbic Acid Oxidase
ACC	1-aminocyclopropane-1-1 carboxylic acid
ACNFP	Advisory Committee on Novel Foods and Processes
ACS	American Chemical Society
ADA	Azodicarbonmide
ADF	Acid Detergent Fiber
ADH	Antidiuretic Hormone
ADI	Acceptable Daily Intake
ADPI	American Dry Products Institute
AECA	Aroma Extract Concentration Analysis
AFLP	Amplified Fragment Length Polymorphism
AGP	Alpha 1-acid glycoprotein
AI	Adequate Intake
AICSIP	All India Co-ordinated Sorghum Improvement Project
AIDS	Acquired Immunodeficiency Syndrome
ALP	Alkaline Phosphatase
ALV	Available Lysine Value
AMC	Automatic Machinery and Electronics
AMCP	Anhydrous Monocalcium Phosphate
AMD	Age-related Macular Degeneration
AML	Amylose Leaching
AMS	Agricultural Marketing Service

Abbreviation	*Full Form*
ANOVA	Analysis of Variance
AOAC	Association of Official Analytical Chemists
AOM	Active Oxygen Method or Atmospheric Oxygen Metabolizing
APFEDA	Agricultural and Processed Food Export Development Authority
APHIS	Animal & Plant Health Inspection Service
APMC	Agricultural Produce Market Committees
APP	Acute Phase Proteins
ARI	Acute Respiratory Infection
ASAE	American Society of Agricultural Engineers
ASTM	American Society of Testing Materials
ATA	Alimentary Toxic Aleukia
ATP	Adenosine Triphosphate
AW	Water Activity
BAM	Bacteriological Analytical Manual
BATF	Bureau of Alcohol, Tobacco and Firearms
BBT	Bright Beer Tank
BHA	Butylated Hydroxy Anisole
BHT	Butylated Hydroxy Toluene
BIS	Bureau of Indian Standard
BMC	Bone Mineral Content
BMD	Bone Mineral Density
BOD	Biological Oxygen Demand
BOP	Broken Orange Pekoe
BOPE	Broken Orange Pekoe Fannings
BRAP	Bilateral Research Activities Program
BSA	Bovine Serum Albumin
BSE	Bovine Spongiform Encephalopathy
BU	Brabender Units
BV	Biological Value
BVA	Brabender Visco Amylogram

Abbreviation	Full Form
BVO	Brominated Vegetable Oils
CA	Controlled Atmosphere
CAC	Codex Alimentarius Committee or Citric Acid Cycle
CACP	Commission for Agricultural Costs and Prices
CAF	Calcium Activated Factor
CANP	Calcium Activated Neutral Proteinase
CAP	Controlled Atmosphere Packaging
CAPP	Calcium Acid Pyrophosphate
CAS	Controlled Atmosphere Storage
CBER	Center for Biologics Evaluation and Research
CC	Column Chromatography
CCI	Cotton Corporation of India
CCK	Cholecystokinin
CCMP	Cooked Cured-Meat Pigment
CCP	Critical Control Point
CDC	Centers for Disease Control
CDER	Center for Drug Evaluation and Research
CDT	Come-Down Time
CE	Capillary Electrophoresis
CEPCI	The Cashew Nuts Export Promotion Council of India
CER	Carbon dioxide Evolution Rate
CFB	Corrugated Fiber Board
CFAM	Cyclic Fatty Acid Monomers
CFC	Chloro Fluoro Carbon
CFTRI	Central Food Technology Research Institute
CFU	Colony Forming Units
CGC	Canadian Grain Commission
CHARM	Combined Hedonic and Response Measurements
CHD	Coronary Heart Disease
CIP	Clean-in-Place

Abbreviation	Full Form
CLA	Conjugated Linoleic Acid
CM	Carboxymethyl
CM	Chloroform-Methanol
CMC	Carboxy Methyl Cellulose
COD	Chemical Oxygen Demand
COSAMB	Council of State Agricultural Marketing Boards
CPg	Specific heat of grain
CPS	Counts Per Second
CPw	Specific heat of water
CRD	Complete Randomized Design
CROs	Contract Research Organizations
CRP	C-Reactive Protein
CRS	Chinese Restaurant Syndrome
CS	Chemical Score
CSA	Cross-sectional open area
CSIR	Council of Scientific and Industrial Research
CTC	Crush Tear Curl or Charge Transfer Certificate
CUT	Come-Up Time
CVA	Cerebrovascular accident
CVD	Cardio Vascular Disease
CW	Continuous Wave
CWC	Central Ware Housing Corporation
D.B.	Dry bulb
d.b.	Dry weight
DAGs	Diacylglycerols
DAP	Diamino Pimelic Acid
DAS	Diacetone-L-Sorbose
DVE	Diacetyl Tartaric Acid
DATEM	Diacetyl Tartaric Esters of Monoacylglycerols
DBS	Dried Blood Spot

Abbreviation	*Full Form*
DCPD	Di-Calcium Phosphate Dihydrate
DDG	Distillers Dry Grains
DDM	Dialkyl Dihexadecyl Malonate
DDS	Distillers Dry Solubles
DE	Dextrose Equivalent or Degree of Esterification
DE	Diatomaceous Earth
DEAE	Diethylaminoethyl
DEFF	Design Effect Fortified Food
DEFT	Direct Epifluorescent Filter Technique
DEPC	Diethyl Pyrocarbonate
DFD	Dark, Firm and Dry
DHA	Docosahexaenoic Acid
DHA	Dehydroalanine
DHAA	Dehydroascorbic Acid
DHS	Demographic and Health Survey
DISCUS	Distilled Spirits Council of the United States
DLVO	Derijaguin and Landau and Verwey and Overbeek
DM	Degree of Methylation
DMA	Dynamic Mechanical Analysis
DMA	Dimethylamine
DMAPP	Dimethyl allyl Pyrophosphate
DMDC	Dimethyl Dicarbonate
DMI	Directorate of Marketing and Inspection
DMTA	Dynamic Mechanical Thermal Analysis
DNA	Deoxyribonucleic Acid
DO	Dissolved Oxygen
DOPA	Dihydroxy Phenylalanine
DP	Degree of Polymerization or Deep Press or Diastatic Power
DRR	Dough Rate of Reaction
DRV	Daily Reference Values

Abbreviation	Full Form
DS	Degree of Substitution
DSC	Differential Scanning Calorimetry
DSM	Dutch States Mines
DTT	Dithiothreitol
DUS	Distingness University and Stability
Dv	Diffusivity
EA	Emulsion Activity
EAR	Estimated Average Requirement
ECA	Essential Commodity Act
ECD	Electron Capture Detector
ECOST	European Corporation in the field of Scientific and Technical Research
EDB	Ethylene Dibromide
EDCT	Ethylene Dichloride Carbon tetrachloride
EDTA	Ethylenediaminetetraacetic Acid
EFEMA	European Food Emulsifier Manufacture's Association
EGF	Epidermal Growth Factor
EP	Redox Potential
EHEDG	European Hygienic Equipment Design Group
EIA	Enzyme Immuno Assay
EIC	Export Inspection Council
ELISA	Enzyme Linked-Immuno Sorbent Assay
ELMC	Equilibrium Moisture Content
EM	Electromagnetic
EMA	Equilibrium Modified Atmosphere
EMC	Equilibrium Moisture Content
EME	Electro-Magnetic Energy
EMIT	Enzyme Multiplied Immuno Assay Technique
EPA	Environmental Protection Agency
EPA	Eicosapentaenoic Acid

Abbreviation	Full Form
EPA	Esterified Phenolic Acid
EPA	Eicosapentaenoic Acid
EPC	Epicatechin
EPG	Esterified Propoxylated Glycerol
EPI	Expanded Programme on Immunization
EPR	Electron Paramagnetic Resonance
ERH	Equilibrium Relative Humidity
ES	Emulsion Stability
ESADDI	Estimated Safe and Adequate Daily Dietary Intake
ESI	Electrospray Ionization
ESLR	Extended Shelf-Life Refrigerated
ESP	Epithio Specifier Protein
EST	Expressed Sequence Tag
ETO	Ethylene Oxide
EVA	Ethylene Vinyl Acetate
EVAC	Ethylene Vinyl Alcohol Copolymer
FACS	Fluorescence Activated Cell Sorting
FAD	Flavin Adenine Dinucleotide
FAD	Food and Drug
FAME	Fatty Acid Methyl Ester
FAO	Food and Agriculture Organization
FAS	Fetal Alcohol Syndrome
FASEB	Federation of American Societies for Experimental Biology
FCC	Food Chemical Codex
FCI	Food Corporation of India
FDA	Food and Drug Administration
FEMA	Flavour Fragrance Materials Association
FEP	Free Erythrocyte Protoporphyrins
FFA	Free Fatty Acid
FFDCA	Federal Food Drug and Cosmetic Act

Abbreviation	Full Form
FIA	Fluorescence Immuno Assay
FID	Flame Ionization Detector
FIFRA	Federal Insecticide, Fungicide and Rodenticide Act
FM	Freezing-Melting
FMA	Fragrance Materials Association
FNB	Food and Nutrition Board
FOS	Fructooligosaccahrides
FOSHU	Foods for Specific Health Use
FPA	Free Phenolic Acid
FPC	Fish Protein Concentrate
FPO	Food Product Order
FPP	Famesyl Pyrophosphate
FRM	Fat-reduced Meat
FSIS	Food Safety and Inspection Service
FSO	Food Safety Objective
FTIR	Fourier Transformation Infrared Spectroscopy
GAP	Good Agricultural Practices
GC	Gas Chromatography
GC-MS	Gas Chromatography-Mass Spectroscopy
GCOH	Gas Chromatography Olfactometry of Headspace
G-CSF	Granulocyte Colony-Stimulating Factor
GGPP	Geranyl Geranyl Pyrophosphate
GHP	Good Hygiene Practice
GL	Glycolipids
GLC	Gas Liquid Chromatography
GLP	Good Laboratory Practice
GMO	Genetically Modified Organisms
GMP	Good Manufacturing Practice
GMPR	Good Manufacturing Practice Regulations
GOT	Glutamate-Oxalacetate Transaminase

Abbreviation	Full Form
GPP	Geranyl Pyrophosphate
GPT	Glutamate-Pyruvate Transaminase
GPV	Gross Protein Value
GRAS	Generally Recognized as Safe
GRE	Glucocorticoid Response Element
GSP	Granular Stationary Phase
GTF	Glucose Tolerance Factor
Gy	Gray
H	Humidity
HAA	Heterocylic Aromatic Amines
HACCP	Hazard Analysis and Critical Control Points
HADY	High Active Dry Yeast
HART	Hybrid Arrested Translation
HAZOP	Hazard and Operability
Hb	Hemoglobin
HCN	Hydrogen Cyanide
HDA	Homochiral Derivatizing Agents
HDI	Human Development Index
HDL	High Density Lipoproteins
HDP	Heat Pump Dehumidifier
HEMF	4-hydroxy-2-(or 5-) ethyl-5-(or 2-) methyl-3 (2H) furnone
HFCS	High-Fructose Corn Syrups
HH	Household
HHS	Health and Human Services
HIC	Hydrophilic Interaction Column
HIV	Human Imuno-deficiency Virus
HKI	Helen Keller International
HLB	Hydrophilic Lipophilic Balance
HM	High-Methoxyl
HMF	Hydroxy Methyl Furaldehyde

Abbreviation	Full Form
HMF	Hydroxymethyl Furfural
HMFP	High Moisture Fruit Products
HMG CoA	Hydroxy Beta-methyl Glutaryl CoA
HMP	Hexose Monophosphate Pathway
HMW	High Molecular Weight
Ho	Humidity of atmospheric air, $kg/hr\ m^2$
HOCl	Hypochlorous Acid
HPC	Hydroxypropyl Cellulose
HPCE	High Performance Capillary Electrophoresis
HPLC	High Pressure/Performance Liquid Chromatography
HPLC	High-Pressure Liquid Chromatography
HPMC	Hydroxypropyl Methyl Cellulose
HRGC	High-Resolution Gas Chromatography
HSH	Hydrogenated Starch Hydrolysates
HTC	Hard-to-Cook
HTS	High Throughput Screening
HTST	High Temperature Short Time
IADY	Instant Active Dry Yeast
IARI	Indian Agricultural Research Institute
IBC	Iodine Binding Capacity
IBPA	Insoluble Bound Phenolic Acid
IC50	Inhibitory Concentration
ICAR	Indian Council of Agricultural Research
ICCIDD	International Council for the Control of Iodine Deficiency Disorders
ICUMSA	International Commission for Uniform Methods of Sugar Analysis
ID	Iron Deficiency
IDA	Iron Deficiency Anemia
IDD	Iodine Deficiency Disorders
IDF	International Dairy Federation

Abbreviation	Full Form
IDF	Insoluble Dietary Fiber
IDL	Intermediate Density Lipoprotein
IEF	Isoelectric Focusing
IF	Intrinsic Factor
IFS	International Foundation for Science
IFTM	Instron Food Testing Machine
IGF	Insulin like Growth Factor
ILPS	International Lecithin and Phospholipids Society
ILSI	International Life Science Institute
IMF	Intermediate Moisture Foods
IMMPaCP	International Micronutrient Malnutrition Prevention and Control Program
IMP	Industrial Membrane Process
IMS	Immuno Magnetic Separation
INACG	International Nutritional Anemia Consultative Group
INCAP	Institute of Nutrition of Central America and Panama
IOM	Institute of Medicine
IPNSs	Integrated Plant Nutrition Systems
IPRs	Intellectual Property Rights
IQF	Individually Quick Frozen
IR	Infra Red
IRDA	Insurance Regularity & Development Authority
ISFE	International Society of Food Engineering
ISI	Indian Standards Institution
ISMA	Industrial Scientific and Medical Applications
ISOS	International Organization for Standards
ISTA	International Seed Testing Association
IU	International Unit
IV	Iodine Value
IVACG	International Vitamin A Consultative Group

Abbreviation	Full Form
IVPD	*In Vitro* Protein Digestibility
JCI	Jute Corporation of India
K	Constant
K	Drying constant, 1/hr
KAP	Knowledge, Attitudes, and Practices
kDa	Kilo Dalton
KI	Potassium Iodide
KV	Kilo Volt
LAB	Lactic acid bacteria
LBG	Locust Bean Gum
LBW	Low Birth Weight
LCD	Liquid Crystal Display
LCE	Low-Cost Extrusion Cooking
LCFA	Long-Chain Fatty Acids
LCP	Liquid Cyclone Process
LCR	Ligase Chain Reaction
LDL	Low Density Lipoproteins
LEC	Low-cost Extrusion Cookers
LF	Lactoferrin
LH	Low Hydroxy
LIFDCs	Low-Income Food Deficit Countries
LISA	Low Input Sustainable Agriculture
LM	Low-Methoxyl
LMP	Low Methoxy Pectin
LMW	Low Molecular Weight
LNG	Liquefied Natural Gas
LOAEL	Lowest Observed Advance Effect Level
LOEL	Lowest Observed Effect Level
LOX	Lipoxygenase
LP	Lactoperoxidase

Abbreviation	Full Form
LP	Lacto Peroxidase or Lipid Peroxidase
LPC	Leaf Protein Concentrates
LSE	Lant Stanol Esters
LTI	Lysine Tri-isocyanate
LUVs	Large Unilamellar Vesicles
LYC	Lycopene Cyclase
M	Molarity
MA	Modified Atmosphere
MAGs	Monoacylglycerols
MAHPD	Modified Atmosphere Heat Pump Dehumidifier
MALDITOF	Matrix-assisted Laser-desorption Ionization Time-of-Flight
MAP	Modified Atmosphere Packaging
MAPIT	Micronutrient Action Plan Instructional Tool
MAS	Monoacetone-L-Sorbose
Mbo	Oxymyoglobin
MC	Methyl Cellulose
MCC	Microcrystalline Cellulose
MCH-PCR	Magnetic Capture Hybridization Polymer Chain Reactions
MCP	1-methyl cyclopropene
MCP	Mono Calcium Phosphate
MCTs	Medium Chain Triacylglycerols
MDA	Malondialdehyde
MDSC	Modulated Differential Scanning Calorimetry
MetMb	Metmyoglobin
MF	Modified Filtration or Microfiltration
MFGM	Milk Fat Globule Membrane
MHP	Modified Humidity Packaging
MI	Micronutrient Initiative
MICS	Multiple Indicator Cluster Survey
MIG	Mercury-in-glass

Abbreviation	Full Form
MLVs	Multilamellar Vesicles
MM	Micro-nutrient Malnutrition
MPC	Milk Protein Concentrates
MPEDA	Marine Products Export Development Authority
MRI	Magnetic Resonance Imaging
MPL	Maximum Pesticide Residue Limits
MRP	Maximum Retail Price
MS	Methyl Sulphonate
MS	Moles of Substitution or Mass Spectrometry
MSG	Mono Sodium Glutamate
MSR	Multistage Recycle
MTD	Maximum Tolerated Dose
MUFA	Monounsaturated Fatty Acid
MVS	Machine Vision System
MWCO	Molecular Weight Cutoff
MWM	Molecular Weight Marker
N	Normality
NA	Not Analyzed
NACMCF	National Advisory Committee on the Microbiology Criteria for Foods
NACMF	National Agricultural Cooperative Marketing Federation
NAD	Nicotine Amide Dinucleotide
NAFED	National Agricultural Finance and Economical Development
NAIP	National Agricultural Innovative Project
NAS	National Academy of Sciences
NASBA	Nucleic Acid Sequence-based Amplification
NATP	National Agricultural Technology Project
NBI	Nitrogen Balance Index
NCA	N-carboxy-alpha-Amino Acid Anhydride
NCCF	National Consumers Cooperative Federation

Abbreviation	Full Form
NCDC	National Cooperative Development Corporation
NCHS	National Center for Health Statistics
NCTGF	National Cooperative Tobacco Growers Federation
ND	Not Detected
NDDB	National Dairy Development Board
NDF	Neutral Detergent Fiber
NDGA	Nordihydro Guaiaretic Acid
NDMA	N-nitrosodimethylamine
NDPE	Net Dietary Protein Energy
NDPV	Net Dietary Protein Value
NDR	Neutral Detergent Residue
NDUS	Novelty Distingness, Uniformity and Stability
NF	Nanofiltration
NFC	Not From Concentrate
NFDM	Non Fat Dry Milk
NFE	Nitrogen Free Extract
NHB	National Horticulture Board
NIF	Nasal Impact Frequency
NIH	National Institute of Health
NIN	National Institute of Nutrition
NST	Nitrogen Solubility Index
NISCAIR	National Institute of Science Communication and Information Resources
NLEA	Nutrition Labeling and Education Act
NMN	Nicotinamide Mononucleotide
NMR	Nuclear Magnetic Resonance
NNMB	National Nutrition Monitoring Bureau
NOAEL	No Observed Advance Effect Level
NOEL	No Observed Effect Level
NDPAGE	Non-Denaturing Polyacrylamide Gel-Electrophoresis

Abbreviation	Full Form
NPN	Non-Protein Nitrogen
NPR	Net Protein Ratio
NPU	Net Protein Utilization
NPV	Net Protein Value
NR	Not Recorded or Never-Ripe
NRCS	National Research Centre for Sorghum
NSI	Nitrogen Solubility Index
NSP	Non-Starch Polysaccharides
NSP	Non-Storage Protein
NVMCE	Non-Volatile Methylene Chloride Extract
OIE	Office International Epizootics
OMFs	Oscillating Magnetic Fields
ONC	Ocean Nutrition Canada
OP	Orange Pekoe
ORP	Oxidation-Reduction Potential
OSHA	Occupational Safety and Health Administration
OUR	Oxygen Uptake Rate
P	Phosphorus
PA	Phytic Acid or Polyamides
PABA	Para-Amino Benzoic Acid
PACMS	Primary Agricultural Cooperative Marketing Societies
PAHs	Polycyclic Aromatic Hydrocarbons
PAL	Phenylalanine Ammonialyase
PAO	Palm Acid Oil
PATH	Program for Technology in Health
PBB	Polybrominated Biphenyls
PC	Polycarbonate
PCA	Perchloric Acid
PCB	Polychlorinated Biphenyls
PCCMP	Powdered Cooked Cured-Meat Pigment

Abbreviation	Full Form
PCM	Protein Calorie Malnutrition
PCMB	Para-Chloro-Mercuric Benzoate
PDA	Photodiode Array Detection or Potato Dextrose Agar
PDAs	Personal Digital Assistants
PDCAAS	Protein Digestibility-Corrected Amino Acid Score
PDCB	Partially Defatted Chopped Beef
PDI	Protein Dispersibility Index
PDS	Phytoene Desaturase
PE	Polyethylene
PEF	Pulsed Electric Field
PEM	Protein Energy Malnutrition
PER	Protein Efficiency Ratio
PET	Polyethylene Terephthalate
PF	Pekoe Fanning's
PFA	Prevention of Food Adulteration
PFAD	Palm Fatty Acid Distillate
PFDA	Pure Food and Drug Act
PG	Propyl Gallate
PGE	Polyglycerol Esters
PHT	Post Harvest Technology
PI	Protein Isolate
PIC	Paired Ion Chromatography
PIT	Phase Inversion Temperature
PKC	Palm Kernel Cake
PKOL	Palm Kernel Olein
PL	Phospholipids
PME	Pectin Methyl Esterase
PMO	Pasteurized Milk Ordinance
POD	Peroxidase
ppm	Parts Per Million

Abbreviation	**Full Form**
PPO	Polyphenol Oxidase
PPO	Propylene Oxide
PPPP	Phytoene Pyrophosphate
PPS	Probability Proportionate to Size
PR	Protein Rating
PRE	Protein Retention Efficiency
PS	Protein Score
PS	Polystyrene
PSE	Pale, Soft and Exudative
PSU	Primary Sampling Unit
PSY	Phytoenesynthase
PT	Press-Twist
PTKs	Protein Tyrosine Kinases
PU	Pasteurization Unit
PUFAs	Poly Unsaturated Fatty Acids
PV	Peroxide Value
PVC	Polyvinyl Chloride
PVDC	Polyvinylidene Chloride
PVP	Polyvinyl Pyrrolidone
PVPP	Polyvinylpyrolidone
PW	Propanol-Water
QA	Quality Assurance
QAE	Quaternary Aminoethyl
QC	Quality Control
QPM	Quality Protein Maize
QAC	Quaternary Ammonium Compounds
RA	Risk Analysis
RAPD	Randomly Amplified Polymorphic DNA
RAR	Retinoic Acid Receptor
RBP	Retinol Binding Protein

Abbreviation	Full Form
RCA	Rolling Circle Amplification
RD	Registered Dietitian
RDA	Recommended Dietary Allowance
RDI	Recommended Dietary Intake
RDI	Reference Daily Intake
Rf	Response Factor
RFLP	Restricted Fragment Length Polymorphism
RFLP-PCR	Restricted Fragment Length Polymorphism-Polymerase Chain Reaction
RH	Relative Humidity
RI	Refractive Index
RIA	Radio Immuno Assay
RIAU	Research Institutions and Agricultural Universities
RNA	Ribonucleic Acid
RNI	Recommended Nutrient Intakes
RO	Reverse Osmosis
ROS	Reactive Oxygen Species
RQ	Respiratory Quotient
RSV	Respiratory Syncytial Virus
RTD	Resistance Temperature Device
RTE	Ready-to-Eat
SALP	Sodium Aluminum Phosphate
SAM	S-adenosyl-L-methionine
SAMB	State Agricultural Marketing Boards
SAPP	Sodium Acid Pyrophosphate
SAS	Statistical Analytical System
SAS	Sodium Aluminum Sulfate
SCC	Somatic Cell Count
SCCMO	Special Commodity Cooperative Marketing Organizations

Abbreviation	Full Form
SCF	Supercritical Fluid
SCFA	Short-Chain Fatty Acids
SCMF	State Cooperative Marketing Federation
SCP	Single Cell Proteins
SD	Standard Deviation
SDA	Strand Displacement Amplification
SDAM	State Directorate of Agricultural Marketing
SDF	Soluble Dietary Fiber
SDS-PAGE	Sodium Dodecylsulphate Polyacrylamide Gel-Electrophoresis
SEM	Scanning Electron Micrograph/Microscopy
SEPC	Silk Export Promotion Council
SF	Swelling Factor
SF	Serum Ferritin
SFDA	Small Farm Development Agency
SFAEs	Sucrose Fatty Acid Esters
SFAP	Sucrose Fatty Acid Polyester
SGA	Small for Gestational Age
SGE	Starch Gel Electrophoresis
SH	Sulphydryl
SHMP	Sodium Hexametaphosphate
SIP	Sterilization-in-Place
SMB	Simulated Moving Bed
SMER	Specific Moisture Extraction Rate
SMFs	Static Magnetic Fields
SMUF	Simulated Milk Ultrafilterate
SNIF	Surface of Nasal Impact Frequency
SO	Soybean Oil
SOD	Super Oxide Dismutase
SOPP	Sodium *O*-Phenylphenate

Abbreviation	Full Form
SOPs	Standard Operating Procedures
SP	Sulphoproyl
SP	Storage Protein
SPC	Standard Plate Count
SPS	Sanitary and Phytosanitary
SSL	Sodium Stearoyl Lactylate
SSLL	Sodium Stearoyl Lactoyl Lactate
SSP	Shelf-Stable Products
STC	State Trading Ccorporation
SPT	Standard Pressure and Temperature
SUVs	Small Unilamellar Vesicles
SWCS	State Warehousing Corporations
SWM	Standards of Weight and Measures
TAC	Trialkoxycitrate
TAG	Triacylglycerol
TAGs	Triacylglycerols
TATCA	Trialkoxy Tricarballylate
TBA	Thio Barbituric Acid
TBHQ	Tetra Butylated Hydro Quinone
TBT	Technical Barriers to Trade
TCA	Trichloro Acetic Acid
TCA	Trichloroanisol/Trichloroacetic acid/Tricarboxylic acid
TCD	Thermal Conductivity Detector
TD	True Digestibility
TDF	Total Dietary Fiber
TDI	Tolerable Daily Intake
TEF	Toxic Equivalency Factor
TFA	Trifluoro Acetic Acid
TFA	Transfatty Acids

Abbreviation	Full Form
TFC	Thin Film Composite
TFR	Tempeh Fish and Rice
TfR	Transerrin Receptor
THBP	2,4,5-Trihydroxy Butyrophenone
TLC	Thin Layer Chromatography
TMA	Transcription-Mediated Amplification
TMA	Thermo-Mechanical Analysis
TMA	Trimethylamine
TMAO	Trimethylamine Oxide
TMCT	Thermal Mechanical Compression Test
TNF	Tumor Necrosis Factor
TOFMS	Time of Flight Mass Spectrometry
TOS	Total Organic Solids
TQM	Total Quality Management
TCMF	Tribal Cooperative Marketing Federation
T-RNA	Tetrahymena Relative Nutritive Value
TS	Total Solids
TSAI	Transition State Analog Inhibitors
TSH	Thyroid-Stimulating Hormone
TSNS	Ten State Nutrition Survey
TSP	Texturized Soya Proteins
TTB	Tax and Trade Bureau
TTI	Time-Temperature Integrator/Time-Temperature Indicator
TTT	Time-Temperature-Tolerance
TVP	Textured Vegetable Proteins
TWT	Traveling Wave Tube
UF	Ultra Filtration
UHT	Ultra High Temperature
UI	Urinary Iodine
ULO	Ultra-Low Oxygen
ULV	Unilamellar Vesicles

Abbreviation	Full Form
UNICEF	United Nations International Children Fund
UNICEF	United Nations Children's Fund
UNU	United Nations Union
UP	Utilizable Proteins
USDA	United State Department of Agriculture
USFDA	United State Food and Drug Administration
USI	Universal Salt Iodization
USS	United States Standards
UTLIEF	Ultra Thin Layer Isoelectric Focusing
UV	Ultra Violet
V/V	Volume-to-Volume
VAD	Vitamin A Deficiency
VFA	Volatile Fatty Acids
VFD	Variable Frequency Drive
VFMPT	Vacuum Freezing Multiple-Phase Transformation
VLCD	Very Low Calorie Diet
VLDL	Very Low Density Lipoproteins
VP	Vacuum Packaging
VSP	Vacuum Skin Packaging
VVR	Vessel Volume Ratio
WBT	Wet Bulb Temperature
W/V	Weight-to-Volume
W/W	Weight-to-Weight
WARC	World Administrative Radio Conference
WBP	Water Binding Potential
WFI	Water-for-Injection
WHC	Water Holding Capacity
WHO	World Health Organization
WLF	Williams-Landel-Ferry
WOF	Warm-over Flavour
WRA	Women of Reproductive Age

Abbreviation	Full Form
WSC	World Summit for Children
WSSN	World Standards Services Network
WTO	World Trade Organization
XME	Xenobiotic Metabolising Enzymes
XOD	Xanthine Oxidase
XRE	Xnobiotic Response Element
Z	Z-score
ZDS	Zeta-Carotene Desaturase

3

TERMINOLOGY

Term	Terminology
Abnormal behaviour	Behaviour that is characterized by the convergence of the four criteria of relative infrequency, social deviance, impaired social functioning and personal distress.
Absolute humidity	It is the amount of water vapour actually present in the air.
Absolute pressure	It is pressure above absolute zero pressure.
Absolute threshold	The minimum amount of stimulation necessary to produce a sensation.
Absolute viscosity	Fluid which has constant consistency value if static pressure and temperature are fixed such consistency termed as absolute viscosity.
Accommodation	The process by which schemata are revised completely or developed further and made more complex to fit new incoming information.
Acetyl value	The number of milligrams of potassium hydroxide required to neutralize the acetic acid liberated by the hydrolysis of 1 g of the acetylated fat or oil.
Achievement	Performance of children in mid-year and inspire examinations.
Acid rain	Presence of excess acid in the rainwater.
Acid value	The number of milligrams of potassium hydroxide required to neutralize the free fatty acids in 1 g of fat or oil.
Acidophilus milk	Type of fermented milk produced by development of a culture of *Lactobacillus acidophilus* in milk.
Acquisition	During conditioning, the process by which an organism learns a new response.
Actinomycetes	Aerobic, gram positive bacteria that forms branching filaments of hyphae and sexual spores.

Term	Terminology
Active absorption	Absorption in which a carrier is used and ATP energy is expended.
Active listening	In humanistic psychotherapy, the process by which the therapist tries to grasp both the content of what the client is saying and the feeling behind it.
Active oxygen method (AOM)	It involves maintenance of the sample at 97.8°C while air is continuously bubbled through it at a constant rate. The time required to obtain a specific peroxide value is determined.
Adaptation	The process of changing behaviour to fit changing circumstances.
Additives	Chemical substance added to food during processing or packaging to improve their sensory properties, textural properties and extend their shelf-life.
Adenosine triphosphate (ATP)	The main energy currency for cells. ATP energy is used to promote ion pumping, enzyme activity, and muscular contraction.
ADF (acid detergent fiber)	A fraction of crude fiber consisting mostly of lignocellulose.
Adhesiveness	The energy required to overcome attractive forces between the food and any surface it is in contact with.
Adjuvant	Substance which increase immunity response to an antigen.
Adolescence	The period that begins with the onset of puberty and ends somewhere around age 18 or 19 years.
Adsorption	The tendency of a molecule to attach itself to a finely.
Adult	An individual who is capable of assuming the responsibility of daily work and committed love.
Advertisement	Advertisement is a mass and paid communication of good services or ideas by an identified sponsor.
Aerobic fermentation	The fermentation carried out in the presence of air/oxygen is known as aerobic fermentation.

Term	Terminology
Affinity chromatography	Separation of biological mixture based on highly specific biological interaction like receptor and ligand.
Aflatoxin	Toxin produced in groundnut due to mold, *Aspergillus flvus*.
Aftertaste	The experience that, under certain conditions, follows removal of the taste stimulus; it may be continuous with the primary experience or may follow as a different quality after a period during which swallowing, saliva, dilution, and other influence may have affected the stimulus substance. The result of the persistence of a flavour note, particularly after swallowing.
Ageusia	Lack or impairment of sensitivity to taste stimuli.
Agnosia	Inability to recognize sensations; may be primarily in one sense, e.g., olfactory agnosia.
Agricultural marketing	Human activity directed at satisfying the needs and wants through exchange process. It can be also depended as comprising of all activities involved in supply of farm inputs to the farmers and movement of agricultural product from the farms to the consumers.
Air conditioning	Simultaneous control of temperature, humidity, cleanliness and air motion.
Alcohol yield	With corn containing 60% starch, distillers traditionally obtain 19-19.7 L (5.0-5.2 proof gallons)/0.03 m (bushel). Theoretical yields as liters of absolute alcohol/100 kg of starch are, stoichiometric 72.0 L (100%), maximum level 68.4 L (95%), industrial standards 62-65 L (86-90%). For the maximum level, 5% or more of carbon substrate is consumed by yeast growth and by-product formation.
Alcoholism	Alcohol abuse characterized by psychological dependence, physical dependence and impaired social functioning.
Alkali refining or neutralization	A process in which the oil is heated with a calculated amount of sodium/potassium hydroxide to saponify and remove the free fatty acids from the crude oil.

Term	Terminology
Alpha (a) helix	A spiral shape constituting one form of the secondary structure of proteins, arising from a specific hydrogen-bonding structure.
Alveolvs	Smallest drogans involved in biosynthesis of milk, having lumen and provided with myoephithelial secretory cells.
Amino acid	An organic molecule possessing both carboxyl and amino groups. Amino acids serve as the monomers of proteins.
Amnesia	A psychological disorder in which memories for certain events are unavailable to recall.
Amylase	Starch digesting enzymes from the salivary glands or pancreas.
Amylopectin	A digestible branched-chain polysaccharide made of glucose units.
Amylose	A digestible straight-chain polysaccharide made of glucose units.
Analgesic	Any agent that produces insensitivity to pain without loss of consciousness.
Anemia	A condition in which the haemoglobin content of the blood is lower than normal.
Anesthesia	Loss of sensation with or without loss of consciousness.
Angle of Nip	Angle formed by the tangents to the roll faces at the point of contact between particle and the rolls.
Angle of repose	When a grain mass is arranged in a heap, the angle between its horizontal base and the slope of cone formed is called angle of repose.
Animal fats	It mostly contains large amounts of C16 and C18 fatty acids (oleic and linoleic acids).
Anisidine value	A measure of oxidation of fats/oils beyond peroxide stage. With aldehydes, anisidine forms compounds (Schiff's base), which show a strong absorbance at 350 nm.

Term	Terminology
	In the presence of acetic acid, p-anisidine reacts with aldehydes producing a yellowish colour. The molar absorbance at 350 nm increases if the aldehyde contains a double bond conjugated to the carbonyl double bond. Thus the anisidine value is mainly a measure of 2-akenals. An expression termed the Totox or oxidation value (OV), which is equivalent to 2X peroxide value + anisidine value, has been suggested for the assessment of oxidation in oils.
Annealing	It is the process in which glass is heated up to 1000 F for 15 minutes and slowly cooled to impart stress and crack resistance is called annealing.
	Term annealing generally refers to removal of stress annealing temperature or point being defined as temperature at which stresses in glass are relived in few minutes.
Anorexia nervosa	An eating disorder involving a psychological loss of appetite and self-starvation, resulting in part from a distorted body image and various social pressures associated with puberty.
Anorexia	It is the term used to describe the lack of appetite for food.
Anosmia	Inability to smell, either totally or a particular substance or group of substances.
Antagonist	It is a chemical substance that counteracts the effects of another chemical substance.
Anterograde amnesia	A memory disorder in which humans lose the ability to store information.
Antetaste	A prior taste, or foretaste, usually of short duration, preceding the main taste or flavour characteristic.
Antherosclerosis	It is a disease in which the walls of the arteries become narrowed as a result of the deposition of lipid containing material.
Anthropometry	Measurement of physical and gross composition.
Antibiotic	A chemical agents produced by one organism that inhibits or harmful to other organism.

Term	Terminology
	Antibiotics are chemical substances secreted by some micro-organisms which inhibit the growth and development of other microbes.
Anticaking Agent or Free Flow Agent	Substance added to finely, powdered or crystalline food product to prevent caking, lumping or agglomeration.
Anticaking agents	Chemical substance used to avoid formation of cakes during storage in powdered food.
Antifoaming agents	An agent that reduces foaming often caused by the presence of dissolved proteins and other stabilizers.
Antimicrobial Agent	Substance used to preserve food by preventing growth of microorganisms and subsequent spoilage, including fungistats, mold and rope inhibitors. Also includes antimicorobial agent, antimyotic agents, preservatives and mold preventing agents (indirect additives).
	These are chemical preservatives with anti microbial properties and play an important role in preventing spoilage and assuring safety of foods.
Anti-nutritional factors	Anti-nutritional factors are the factors which affect nutritive value of foods, these are naturally occurring toxicants present in legumes, and fauvism is haemolytic anaemia.
Antioxidants	Substances that retard the oxidative rancidity and thereby improve the shelf life of fats e.g. propyl gallate (PG), octyl gallate (OG), butylated hydroxyanisole (BHA), butylated hydroxy toluene (BHT), tertiary butyl hydroquinone (TBHQ) and tocopherol.
	The substances, which are responsible for the control of free radical-mediated lipid oxidation, are known as antioxidants.
Anxiety hierarchy	A rank ordering of anxiety-provoking situations used in systematic desensitization.
AOM (active oxygen method) time	The time required to obtain a predetermined peroxide value when washed air is bubbled through a fat sample held at 97.8 °C.

Term	Terminology
Aphagia	It means the loss of the power of swallowing.
Aphasia	Impairment in the ability to speak or to understand spoken language.
Aroma	The fragrance or odor of food, perceived by the nose by sniffing. In wines, the aroma refers to odors derived from the variety of grape, e.g., Muscat aroma. It is the overall odor impression as perceived by the nasal cavity.
Arousal	An internal state of excitement or tension. Drive theorists believe that a state of arousal demands the reduction or elimination of tension.
Asepsis	Keeping out micro-organisms.
Aseptic packaging	The technique in which food is sterilize the package and fill in the previously sterilized container and packed in it under sterile environment is known as aseptic packaging.
Aspartame	A sweetener made of two amino acids and methanol; it is 200 times sweeter than sucrose.
Aspiration	A process of cleaning grains by using a large volume of air through grain layer to separate lighter particles.
	It is the process of separation of husk or hull by use of air currents.
Assignment problem	Scheduling jobs to person at the least cost so that one person gets one and only one job is called as assignment problem.
Astringency	Astringency is a taste-related phenomenon, perceived as a dry feeling in the mouth along with a coarse puckering of the oral tissue. Astringency usually involves the association of tannins or polyphenols with proteins in the saliva to form precipitates or aggregates.
Attention	The process that determines which sensations will be perceived; *i.e.,* which information will be transferred from sensory to short-term memory.
Auditory nerve	Axons from neurons within the cochlea that conduct neural impulses to the auditory areas of the brain.

Term	Terminology
Autosmia	Disorder of the sense of smell in which odors are perceived when none are present.
Autosomes	Chromosome present in same number in both sexes of any species.
Autotrophic nutrition	A mode of obtaining organic food molecules without eating other organisms. Autotrophus use energy from the sun or from the oxidation of inorganic substance to make organic molecules from inorganic ones.
Autoxidation	The breakdown of fatty acids into peroxides and hydroperoxides upon addition of molecular oxygen across the double bonds during the storage of oil.
Auxotrophs	Bacterial mutant that require one or more growth factors that wild type or phototropic can synthesis.
Axon	The portion of a neuron that transmits neural impulses away from the soma toward the synapse.
Azeotrope	Mixture of two or more refrigerants behaves like a single compound having fixed boiling point corresponding to given pressure.
Backset	Backset is the screened aqueous by-product from distillation. It is recycled and added to the cooked grain mash prior to fermentation.
Balling	Balling is a measure of the sugar concentration in a grain mash, expressed in degrees. It approximates percent by weight of the sugar in solution.
Basal metabolic rate (BMR)	The minimal number of kilocalories a resting animal requires to fuel itself for a given time.
Beer Lambert's law	It states that the absorbance of a solution is directly proportional to the solutions concentration.
Beer	Beer is the alcoholic product arising from the yeast fermentation of saccharified grain mash. It may or may not include stillage from a previous fermentation/distillation.
Behaviour	Any activity that can be observed recorded and measured.

Term	Terminology
Beriberi	The thiamin deficiency disorder characterized by muscle weakness, loss of appetite, nerve degeneration and sometimes edema.
Beta pleated sheet	A zigzag shape constituting one form of the secondary structure of proteins formed of hydrogen bonds between polypeptide segments running in opposite directions.
Betalins	There are the groups of pigments containing betayanins and Betaxantin (yellow colour) also called as nitrogen containing anthocyanins.
Beverage	It is a fruit juice considerably altered in composition before consumption. It may be diluted before it is served.
Biennial	These are defined as plant which requires two years or at least two growing seasons.
Bio-functional membrane	Used in principle of bio sensors for molecules recognition in the process.
Bioavailability	The degree to which the amount of an ingested nutrient actually gets absorbed and so is available to the body.
Biochemical engineering	Biochemical engineering is concerned with conducting biological processes on an industrial scale, providing the link between biology and chemical engineering.
Biochemical marker	Biochemical markers are the proteins produced by gene expression can be isolated and identified by electrophoresis and staining.
Biological value	Biological value of protein is the body's ability to retain protein absorbed from a food.
Biomagnifications	Accumulation and transfer within food chain or ecosystem.
Bitter pit	Small brown dry area disfigures the flesh location below the skin.
Bitter taste	Some have speculated that bitter taste serves as a detergent from poisonous foods; however, we enjoy many bitter foods, e.g., coffee. Many different types of molecules produce a bitter taste including

Term	Terminology
	divalent cations, alkaloids, and some amino acids. With >30 receptor systems for bitterness, it is the least discriminating of the taste modalities. Bitterness could arguably be broken down into several additional taste classifications. Due to the broad range of chemical structures, multiple analytical approaches may be necessary to analyzing bitter components, through HPLC is often applicable.
Black body	The body which absorbs all the incident radiation and none is reflected or transmitted or one whose absorptivity is 100%.
Blanching	Blanching is an important heating then holding and cooling process to inactivate enzymes.
Bland	Having no distinctive taste or odor property.
Bleaching	A process for the removal of colouring matter from the oil.
Blind spot	A portion of the retina through which the optic nerve exits from the eye and travels to the brain. This area has no rods or cones and is not responsive to light.
Blood brain barrier	A specialized capillary arrangement in the brain that restricts the passage of most substances into the brain, thereby preventing dramatic fluctuations in the brain's environment.
Blood doping	A technique by which an athlete's red blood cell count is increased.
Blood pressure	The hydrostatic force that blood exerts against the wall of a vessel.
Blood	Blood is the fluid which circulates through the body bringing nourishment too and removing waste products from the cells.
Body mass index	Weight (kg) divided by height squared (meter). A value of 30 or greater shows obesity related health risks.
Boiler Water Additive	Substance used in a steam or boiler water system as an anticorrosion agent, to prevent scale or to effect steam purity.

Term	Terminology
Boiler	A combination of apparatus for producing, furnishing or recovering heat together with the apparatus for transferring the heat so made available to the fluid being heated and vaporized.
Boiling point	A liquid boils when its vapour pressure is equal to the external pressure. The boiling point is thus constant for any external pressure. The normal boiling point refers to an external pressure, which is equal to the atmospheric pressure (760 mm Hg), which for water is 100°C.
Bomb calorimeter	An instrument used to determine the Kcalorie content of a food.
Bond	A sharing of electrons, charges or attractions. This links two atoms.
Bonded Whiskey	Bonded whiskey is whiskey stored at least four in wooden containers where the spirits have been in contact with the wood surface. It is unaltered from the original character by the addition or subtraction y substance other than by filtration or chill proofing, is reduced in proof by addition of water to 100° proof (50 vol %) and bottled at 100° proof, and is produced at the same distillery in the same season (January through June or through December).
Bound water	In certain foods some of the water may be bound very closely and very different to separate. Often it is extremely different to separate the water without decomposing other molecules present in the food sample. Water that is attached to organic substances. This is not available for microbial use.
Boyl's law	PV = Constant, P1V1 = P2V2 = P3V3 = Constant.
Brain power	Brain power is characterized by how alert, energetic, and concentrated the brain of a person is in response to a task.
Brainstem	The bottommost protein of the brain, an enlarged extension of the spinal cord consisting of the medulla, pons, midbrain and thalamus.

Term	Terminology
Bran	Corresponding fraction from the last break of grind in milling of wheat or any cereal grains or the brownish covering of cereal grains below the husk, that consists of pericarp, testa and aleurone layer is called bran.
Bread staling	Change in crust and crumb of bread.
Break-even point	The point at which the level of sales just equal to total expenses.
Brown rice	Dehusked rice grain obtained after dehusking paddy. Also called unpolished rice, dehusked rice.
Browning	Browning of foods is due to oxidative or non-oxidative reactions. Oxidative or enzymic browning is a reaction between oxygen and a phenolic substrate catalyzed by polyphenol oxidase. Nonoxidative or nonenzymic browning involves the phenomena of carmelization and or the interaction of proteins or amines with carbohydrates.
	Browning reaction contributes to the aroma, flavour, and colour of many foods such as ready-to-serve cereals, toffees, roasted coffee, malted barley and baked goods.
	Certain enzymatic and non-enzymatic changes occurring in food mostly fruits and vegetable due to oxidation reaction cause change in colour of the food product is known as browning.
Bubble gum	It is confection which can produce bubbles in mouth. It is prepared by using gum, resins and plasticizer with sugar and glucose syrup.
Buffer	Compounds that can take up or release hydrogen ions to maintain a certain pH value in a solution.
Bulk density	Mass of particles occupied by unit volume of bed.
Butter oil	Butter oil is the fat concentrate obtained mainly from butter or cream by the removal of practically all the water and solid non-fat content.
Butter	It is a fat or cream separated from other milk constituents.

Term	Terminology
BV (biological value)	The percent of the absorbed nitrogen actually retained in the body.
Cacogeusia	Persistent or intermittent unpleasant taste in the mouth.
Cacosmia	Perception of persistent or intermittent unpleasant odor.
Cake	A product obtained by baking a leavened and shortened batter containing flour, sugar, egg, milk, flavor and leavening agents.
Calorie-free	A product with fewer than 5 cal per serving.
Calories	A measurement of energy provided by foods or The amount of heat energy required to raise the temperature of 1 g of water $1°C$; the amount of heat energy that 1 g of water releases when it cools by $1°C$. The Calorie (with a capital C), usually used to indicate the energy content of food, is a kilocalorie.
Calorific value	The amount of energy (kcal) produced when one gram of substance is completely oxidized.
Cancer	A condition characterized by uncontrolled growth of abnormal body cells.
Canning of peas	Heat processing in hermetically sealed container.
Canning	It is defined as the preservation of food in hermetically sealed containers which usually implies heat treatment as the principal factor in the prevention of spoilage.
Caramel	Sugar heated beyond their melting point decompose and form a brown mass known as caramel.
Caramelization	Direct heating of carbohydrates, particularly sugars and sugar syrups, produces a complex group of reactions termed Caramelization.
Carbohydrate	A compound containing carbon, hydrogen and oxygen atoms; most are known as sugars, starches and dietary fibers.
Carbon dioxide injury	In CA storages were some times CO_2 concentration goes higher then, the periphery of the internal tissue turns slightly brown initially then deep brown.

Term	Terminology
Cariogenic	A substance often carbohydrate-rich that promotes dental caries.
Carotenes	Pigment substances in plants that can often form vitamin A. Beta-carotene are the most active form.
Casein	It is principle milk proteins, having heterogeneous mixture of various proteins.
	Proteins in milk that form hard curds. These are difficult for infants to digest.
Catabolism	Breaking down compounds.
Catalyst	A compound that speeds reaction rates but is not altered by the reaction.
Cavitations	If the pressure in the suction line is less than the vapour pressure, some of the liquid flashes into vapour or if the liquid contains gases, they may come out of the solution resulting into gas pockets. This phenomenon is known as cavitations.
Celiac disease	Also known as gluten-induced enteropathy. It is caused by an allergy to protein found in wheat, rye, oats and barley. If untreated, it causes a severe flattening of the villi in the intestine, leading to severe malabsorption of nutrients.
Cellulose	A straight-chain polysaccharide of glucose molecules that is undigestible because of the presence of beta carbohydrate bonds.
Centrifugal clarification	The removal of small quantities, a few per cent or les of insoluble solids from a liquid by centrifugal mens.
Centrifugal force	Centrifugal force = mRw2.
Cereals	Cereals are the fruits of cultivated grasses and they belong to family graminae they are monocotyledons.
	Seeds of grass belonging to graminiae family.
CGRE	Combine glass reference electrode used to measure the difference in pH in bioreactor.
Chalky grains	The grains which appear milky/white/opaque called chalky grains.

Term	Terminology
Charles law	V/T - Contstant.
Cheilosis	Cheilosis is the condition characterized by lesions of the lips and at the angles of the mouth, which is caused by a deficiency of the vitamin riboflavin.
Chemesthesis	Chemesthesis in the mouth is the chemical irritation (e.g., pain, heat, cooling) due to stimulation of the trigeminal nerve. Chemesthesis also occurs in other parts of the body including the eyes, nose, and throat. Some examples of chemesthesis are burning from jalapenos, cooling from mint, and pain from carbonation.
Chemical score	The percentage of the ratio of the most limiting amino acid in the test protein to that in the standard (whole egg) protein.
Chilling injury	Commodities (mainly tropical and subtropical) held at temperature above their freezing point and below 5 to 15°C depending on the commodity.
	Exposure of tissues of commodities to temperature below critical level of storage temperature leads to imbalance of metabolism and causes injury to tissue.
Cholecystokinin (CCK)	A hormone that stimulates enzyme release from the pancreas and bile release from the gallbladder.
Cholesterol	A waxy lipid. It has a structure containing multiple chemical rings.
Chromatography	Chromatography can be defined as a primarily as a process separation which is used for the separation of molecular mixture.
Chronic toxicity	Refers to an effect that requires some time to develop toxicity.
Chronic	Long-standing, developing over time; slow to develop or resolve. When referring to disease this indicates that the disease progress, one developed, is slow and tends to remain; a good example is coronary heart disease.

Term	Terminology
Chylomicrons	Dietary fat surrounded by a shell of cholesterol, phospholipids and protein. These are made in the intestine after fat absorption and travel through the lymphatic system to the bloodstream.
Cis isomer	An isomer form seen in compounds with double bonds, such as fat where the hydrogen on both ends of the double bond lie on the same side of that bond.
Cleaning of wheat	Process of removal of impurities from wheat.
Climacteric fruits	Fruits which ripens after harvesting.
Clinching	Loosely covering of lids before exhausting in canning technology.
Cling film wrapping	It is the pretreatment in which the fresh fruits are wrapped in very thin film of polyethylene having a glossy edible adhesive, reduced water vapour transpiration rate, high gas permeability and do not require vents which helps to reduce the physiological loss in weight and increases shelf life.
Clostridium botulinum	A bacterium that can cause a fatal type of food poisoning.
Cloud point	The temperature at which the first permanent cloud, due to the crystallization of high-melting materials, appears in the body of the oil when cooled under standardized conditions.
Cloud test	The time (h) required for the appearance of visible cloud in the salad oil held at 0°C.
Cloying	A taste sensation that stimulates beyond the point of satiation; frequently used to describe overly sweet products.
Coagulation	Random aggregation reactions with denaturation and aggregation reactions where protein-protein interactions predominate over protein-solvent interactions are defined as coagulation.
Coarse dispersion	Dispersion having particle size greater than 0.5 μm is coarse dispersion.
Coccus	Comma shape bacteria.

Term	Terminology
Co-culture	Two different species/genotypic cells are grown on a cell culture.
COD	Chemical oxygen demand is the measure of oxygen consumed during the oxidation of the oxidizable organic matter by a strong oxidizing agent or Amount of O2 required by organic matter in a sample of water for its oxidation by strong chemical oxidant.
Codon	The nitrogen bases present on mRNA and code for a specific amino acid.
Cohesiveness	The strength of the internal bonds in the sample.
Cold chain	It is the provision of lowest safe temperature to the commodities immediately after its harvest till it reaches the consumer.
Cold sterlization	Microbial destruction without the generation of high temperature with the help of U V radiations.
Cold test	The temperature at which the oil becomes solid.
Color or Coloring Adjunct	Substance used to impart, preserve, or enhance the color or shading of a food. Includes colore, fixatives, color-retention agents, etc.
Colostrum	The first fluid secreted by the breast during late pregnancy and the first day after birth. This thick fluid is rich in immune factors and protein.
	It is first milk secreted by mammary/milking glands of all living females after birth of young one. It is rich in immunoglobulins, anti-microbial peptides (lactoferrin and lactoperoxidase) and other bioactive molecules, including growth factors. Human colostrums have much higher concentrations of epidermal growth factor.
Colour blindness	A condition in which individuals lack the ability to discriminate among different wavelengths of light.
Colour constancy	The tendency for a familiar object to be perceived as a constant colour even though the light that it reflects changes the sensation.
Coma	A state of deep unconsciousness below normal sleep.

Term	Terminology
Communication	Communication is a process by which two or more people exchange ideas, facts feelings or impressions in ways that each gains a common understanding of meaning, intent and use of message.
Compatibility	In flavour terminology the ability of one substance to enhance the flavour characteristic of another.
	It is nothing but package neither should change its colour, flavor, foreign test to food material nor should it absorb it from material.
Compensation	The result of interaction of the components in a mixture of stimuli, each component of which is perceived as less intense than it would be alone.
Competence	Ability of recipient cell to take up DNA through changes in cell wall.
	Uptake of free DNA fragments and incorporation into its genome by a prokaryotic cell.
Complete proteins	Proteins that contain ample amounts of all ten essential amino acids (PVT TEAM HALL).
Compressible fluid	When the density of fluid is sensitive to change in temperature and pressure, the fluid is said to be compressible fluid.
Concentrate	This is a fruit juice, which has been concentrated by removal of water either by heat or by vacuum drying.
Conditioning	Process of soaking grains in water to dehydrate them and for even distribution of moisture in grain.
Conduction	It is the transfer of heat between adjacent molecules without appreciable displacement of the particles of the body.
Conformity	Adopting the attitudes and behaviors of other people as a result of real or imagined pressure from others.
Congeners	Congeners are the flavor constituents in beverage spirits that are responsible for its flavor and aroma and that results from the fermentation, distillation, and maturation processes.

Term	Terminology
Constitutive enzymes	These are present in bacterial cells in constant amount regulates of the metabolic state of the organizm.
Constraints	Physical limitations on the set of decision variable.
Contribution	The difference between sales revenue and variable cost is known as contribution.
Control system	When the measuring and controlling instruments after combined so that measurements provide impulses for remote automatic action the result is called a control system.
Convection	Convection is the transfer of heat from one point to another within a fluid, gas or liquid by the mixing of one portion of the fluid with another. The motion of the fluid may be entirely the result of differences in density resulting from the temperature differences. This is known as natural convection. The motion may also be produced by mechanical means as in forced convection.
Convergence	The tendency of a test sample, regardless of quality, to be perceived as similar to prior sample(s); sometimes called the halo effect.
Conversion	Conversion describes the enzymatic starch hydrolysis processes, liquification, and saccharification.
Convolutions	The irregular "hills and valleys" that make up the surface of the cortex in humans and other complex animals.
Convulsion	It is a condition marked by involuntary contraction of the muscles.
Cooking	Cooking is the gelatinization by heat treatment and alpha-amyase liquification, and raw material starch.
Cooling	Cooling sensations occur when certain chemicals contact the nasal or oral tissues and stimulate a specific saporous receptor.
Cooperation	Form of association of people to work together in order to active a particular end.

Term	Terminology
Cordial	It is a sparking, clear sweetened fruit juice from which all pulp and other suspended materials have been completely eliminated. Limejuice cordial is a typical example of cordial.
Cornea	The transparent, domelike, outer covering of the lens and iris that allows light to enter the eye.
Cortical bone	Dense, compact bone that comprises the outer surface and shafts of bone.
Cortisol	A hormone made by the adrenal gland that, among other functions, stimulates the production of glucose from amino acids.
Covalent bond	A union of two atoms formed by the sharing of electrons.
Cream	Cream may be defined as that portion of milk is rich in milk fat.
Critical control point	Identify critical control points in the process at which the potential hazard can be controlled or eliminated.
Critical point in grain drying	It is the point on the drying curve at which the constant rate period ends and falling rate period starts.
Critical speed of mill	Speed at which the small sphere inside the mill just begins to centrifuge.
Crop insurance	Crop insurance scheme was launched to protect the farmers from heavy loses of crop due to rain fall, floods, drought etc.
Crude fiber	What remains of dietary fiber after acid and alkaline treatment? This consists of primary cellulose and lignin.
Crush	Type of fruit beverage which contains at least 25% fruit juice or pulp and 55% TSS.
Crusting	Formation of dry crust on surface of dough due to evaporation of water.
Cryogenic refirgerants	Those are employed for achieving the temperature in the range of 113 to 0°K.
Cryptosmia	Impairment of olfaction by obstruction of the nasal passages.

Term	Terminology
Crystallized fruits/ vegetables	Candied fruits/vegetables covered or coated with crystals of sugar or finely powdered sugar are called crystallized fruits/vegetables.
Curing or Pickling Agent	Substance imparting a unique flavor and/or color to food, usually producing an increase in self-life stability.
Curing	Keeping of the commodity in a field for a week to heal up the wounds and to maintain the moisture content.
Custard	Sweet mixture of egg and milk baked or cooked over hot water
Daily Reference Values (DRV)	Standards of intake for certain parts of a diet, such as carbohydrate, fat, saturated fat, cholesterol, sodium, potassium and dietary fiber, set by FDA for which no U.S. RDA exist. These values are intended to be used for comparing intake of these factors to desirable levels of intake.
Dampness	The paper boards observed moisture during rainy season when humid atmosphere is there the absorption of moisture package become loose its strength is a dampness.
Data	Records of the observations or measurements in an empirical study.
Deaeration	Freshly extracted and screened juices contain appreciable quantity of oxygen which should be removed before packing this process is known as deaeration.
Debentures	It is one of the methods of raising finance by issuing, debentures against security of its. Holders receive fixed interest.
Debittering	The process of removing bitter substances from oils or fats.
Decimal reduction time	Time over which the spore concentration is reduced by ten fold.
Decomposer	Organisms with secrete digestive enzymes to breakdown food into simpler substances.

Term	Terminology
Defatted meal	Meal obtained after extraction of oil, utilization, defatted meal.
Degerming	The process of separating germ or embryo from the seed or kernel.
Degree of milling	The extent to which undesirable parts like husk, bran etc. are separated from paddy.
Degreening	It is the process in which treatment with ethylene under controlled conditions hastens the los of chlorophyll.
Dehulling	Dehulling is the process of removing hull during milling.
Delinting of cotton seed	Cottonseed contains residual fibers after ginning called as lint. The process of removal of these lints in battery machines is known as delinting of cottonseed.
Denaturation	A process in which a protein unravels and loses its native conformation, thereby becoming biologically inactive. Denaturation occurs under extreme conditions of pH, salt concentration, and temperature.
Dental caries	Erosions in the surface of a tooth caused by acids made by bacteria as they metabolize sugars.
Deodorization	A process in which the heated oil is placed in a vaccumized tower and allowed to cascade over steam moving in countercurrent direction, to remove steam-volatile substances contributing to undesirable flavour and odor to the oil.
Depression	An extreme mood of despondency, hopelessness, and lowered self-esteem.
Designer Food	Designer Food is a processes food that is supplemented with food ingredients naturally rich in disease-preventing substances (Ca fortified flour, Low sodium foods).
Desludging	This term is used to describe the removal centrifugal means when the quantities of solids present more than 5 per cent.
Desolventizers	Used for removal of traces of solvent from oil and miscella.

Term	Terminology
Dew point	The temperature at which a given sample of moists air will become saturated and deposit dew.
Dextrin	Partial breakdown products of starch that contain few to many glucose molecules. These appear while starch is being digested into many units of maltose.
Dextrose equivalent	It is total reducing power of glucose syrup expressed on a dry solids basis as dextrose. It is measure of degree of hydrolysis.
Diabetes mellitus	A disease characterized by high blood sugar levels (hyperglycemia), resulting from either an insufficient insulin released by the pancreas or a general inability for insulin to act on certain body cells, such as adipose cells.
Dietary fiber	Substances in food that are not digested by the processes present in the stomach and small intestine.
	Dietary fibre is the indigestible carbohydrate portion of plant foods (including cereals, grains, oilseeds, fruits, vegetables, herbs). It is also sometimes called roughage or simply fibre. It basically includes those plant cell wall components that are not digestible by human or mammaliam digestive enzymes. The term *"dietary fiber"* was coined by Hipsley (1953) referring to "a non-digestable constituents making up the plant cell wall.
Dietary Supplement	A product that contains one or more of the dietary ingredients *i.e.* vitamin, mineral, herb, or other botanical, and amino acid (protein). Includes any possible component of the diet as well as concentrates, constituents, extracts or metabolites of these compounds.
Dietetics	It is a science that deals with the adequacy of diet during normal life cycle and modification required during diseased condition.
Differential pressure	The difference between two pressures is called different pressure.

Term	Terminology
Diffused Reflectance	When light is reflected from a surface evenly at all angles then object look dull is termed as diffused reflectance.
Digestion	The process of breaking down food into molecules small enough for the body to absorb.
Dilatent material	The material which exhibit an increase of apparent viscosity as rate of shear increase.
Direct calorimetry	A method to determine energy use by the body by measuring heat that emanates from the body, usually using an insulated chamber.
Disaccharides	Class of sugars formed by the chemical bonding of two monosaccharides.
Discrimination	In conditioning, a process whereby the organism learns to respond differently to stimuli that are differentially reinforced. In social psychology, the expression of prejudice in hostile behaviour toward members of a particular group.
Distillation	The separation of components of a solution on the basis of their volatilities is known as distillation.
Distortion	The process by which information in memory is altered, making it inaccurate.
DNA Chimera	DNA containing genetic information derived from two different species or genotypes.
DNa polymerase II	It is second member of endonuclease family of enyzme. It remains active on DNa duplexes having gaps.
Dockage	The impurities found in grains such as chaff, stalks, grain dust, other seeds, immature grains are collectively called as dockage.
Docosahexaenoic acid (DHA)	An omega 3 fatty acid with 22 carbons and six carbon-carbon double bonds. DHA is also present in fish oils and also may be synthesized from alpha-linolenic acid.
Documentation	Establish effective record-keeping procedures that document and provide a historical record of the facility's food safety performance.

Term	Terminology
Dominant gene	A member of a gene pair that controls the expression of a physical trait regardless of the nature of the other member of the gene pair.
Doubler	A doubler is a pot still used to redistill whiskey and low wines from a beer still. The low wines are fed into the doubler where they are redistilled by way of steam enclosed in a scroll at the bottom of the still. The bottoms, the organic components remaining at the bottom of the still, are returned to the still to extract the alcohol.
Dough improvers	The chemical substances used to improve viscosity, loaf volume, cellular texture, freshness, extensibility etc. of the bread.
Dough Strengthener	Substance used to modify starch and gluten, thereby producing more stable dough.
Down syndrome	A form of organic retardation resulting from an abnormality in which the child has one more chromosome than normal.
Drug	A substance intended for use in the diagnosis. cure, mitigation, treatment or prevention of disease in man or other animals. A substance other than food intended to affect the structure or any function of the body of man of other animals."
Drying Agent	Substance with moisture-absorbing ability, used to maintain an environment of low moisture.
Drying oil	Oil which when applied as a thin film readily absorbs oxygen from the air and dries to form a relatively tough elastic substance.
Drying	Removal of water (moisture) by using non-conventional energy sources is called as drying.
Dryness fraction of steam	Ratio of the mass of actual dry steam to the mass quantity of wet steam.
Dumping	It is selling goods in foreign country at a price which local producer regards as unfairly low.
Dye	Dyes are any substance that lend colours materials or dye is a food-grade water soluble colorant-FDA.
Dysomia	Difficulty in ability to smell.

Term	Terminology
Ecology	The study of relationship between living and their surrounding.
Economizer	It is device which serves to recover some of the heat being carrying by exhaust gases.
Edema	Edema is the presence of an abnormally large amount of fluid in the tissue spaces of the body.
Edible films and coating	It can be defined as thin layers of edible material applied on foods by wrapping immersing, brushing or spraying in order to offer selective baser against transmission of gases vapours and solutes while also offering mechanical protection.
Effluent	It is a general term for a fluid emitted by source.
Eicosanoids	Hormone like compounds synthesized from polyunsaturated fatty acids. Within this class of compounds are prostaglandins, thromboxanes and leukotrienes.
Eicosapentaenoic acid (EPA)	An omega-3 fatty acid with 20 carbon atoms and five double bonds; present I fish oils and may be synthesized from alpha-linolenic acid [20:5].
Electrocardiogram	A plot of electrical activity of the heart over the cardiac cycle; measured via multiple skin electrodes.
Electrophoresis	It is the migration of colloidal particles through a solution under the influence of an electric field.
Emotion	A state made up of a characteristic subjective experience, a characteristic pattern of physiological arousal, and a characteristic pattern of overt expression.
Emulsification/Emulsion	Operation in which two normally immiscible liquids are intimately mixed or When an immiscible liquid is dispersed as small droplets in another immiscible liquid by mechanical agitation, an emulsion is formed.
Emulsifier or emulsifier Salt	Substance, which modifies surface tension in the component phase of an emulsion to establish a uniform dispersion or emulsion.

Term	Terminology
Emulsion Capacity (EC)	It is the volume of oil (milliliters) that can be emulsified per gram of protein before phase inversion occurs.
	The volume (ml) of liquid emulsified by 1 g of flour or protein.
	The volume of liquid emulsified by 1 g of flour or protein.
Emulsion stability (ES)	It is ratio between volumes of final emulsion x 100 divided by volume of initial emulsion.
Emulsion	Emulsion is a mixture of two immiscible liquids. Milk cream, mayonnaise, salad dressings, ice cream mix, and cake batters are all O/W emulsions. Butter and margarine are W/O emulsions.
Enrichment	Addition of specific amounts of selected nutrients in accordance with a standard identity as defined by the U.S. Food and Drug Administration.
	Enrichment has been used interchangeably with fortification (FAO/WHO, 1994), but in general enrichment can be defined as the restoration of vitamins and minerals lost during processing.
Entrepreneurship	Entrepreneurship can be defined as a creative innovative response to the environment
Environment	Physical, chemical, biotic and cultural conditions and their ramification collectively comprise.
Enzyme	Enzyme is biocatalysts, organic catalyst, it accelerate rate of reaction of It is a biocatalyst which enhances the rate of reaction without under goes itself.
	Proteins that catalyze specific biological reactions with extraordinary catalytic power.
Equilibrium moisture content	It is moisture content of solid in equilibrium with surrounding conditions when there vapour pressure are equal.
Erythroprotein	A protein secreted by the kidneys that enhances red blood cell synthesis and stimulates red blood cell release from bone marrow.

Term	Terminology
Essence recovery	An essence recovery system which employs a "stripping column" to separate the volatile essence from the less volatile water.
Essential amino acids	Amino acids not efficiently synthesized by humans that must therefore be included in the diet. There are ten essential amino acids.
Essential fatty acids	The fatty acids, which are not synthesized in the human body that must therefore be included in the diet. They are linoleic and alpha-linolenic acid.
Essential mineral	It is an inorganic substance which is needed by at least two species of animals for good health
Essential nutrient	A chemical element required for a plant to grow from a seed and complete the life cycle. A nutrient substance that an animal cannot make itself from raw materials but that must be obtained in food in prefabricated form.
Essential oil	The volatile material, derived by a physical process, usually distillation, from odorous plant material of a single botanical form and spices with which it agrees in name and odour.
Ester value	The difference between the saponification value and the acid value.
Esterification	With regards to fats, the process of attaching fatty acids to a glycerol molecule, creating an ester bond. Removing a fatty acid is called deesterification; reattaching a fatty acid is called reesterification.
Ethylene	The only gaseous plant hormone, responsible for fruit ripening, growth inhibition, leaf abscission and aging.
Evisceration	To cut the belly portion of slaughtered animal.
Exoenzyme	These are the enzymes produced by microbial cells outside the cell structure or excreted in fermentation media during metabolism.
Expectancy	One aspect of the cognitive view of conditioning, which holds that the organism comes to expect a einforce.
Expel out	Separation of liquids from solids by the application of compressive forces.

Term	Terminology
Experiment	A study in which a scientist treats an object of study in a specific way and then observes the effects of the treatment.
Expression	Expression is the separation of liquids from solid by the application of compressive forces.
Extra lean	Meat, poultry, seafood, and game meats containing < 5g of fat per serving.
F0 value	Number of minutes to specific temperature required to destroy a specific number of organisms having a specific 'Z' value.
Falling rate period	Period of drying in which the role of drying decreases with time due to slow removal of internal moisture.
False fruit	Fruit is said to be false fruit when it develops collectively along with progressive development of ovary and associated parts.
Famine	A time of massive starvation often associated with crop failures, war and political strife.
Fat analog	A compound that provides food with many of the characteristics of fat, but altered digestibility and altered nutritional value.
Fat barrier	An ingredient system that provide a barrier for products that use fat as a heat exchange medium.
Fat extender	A fat replacement system, containing a proportion of fat combined with other ingredients, designed to optimize the functionality of fat, thus allowing a decrease in the usual amount of fat in the product.
Fat mimetics	A fat replacer that can mimic one or more of the organoleptic and physical functions of fat, and usually requires a high water content to achieve its functionality.
Fat replacer	An ingredient that can be used to replace fat, yielding fewer calories than fat and may or may not provide nutritional value

Term	Terminology
Fat substitute	A synthetic compound, usually having a similar chemical structure to conventional fats and oils, designed to replace some or all of the functions of fat without any energy contribution, and is usually indigestible or unabsorbable.
Fat	A biological compound consisting of three fatty acids linked to one glycerol molecule.
Fat-free	A product containing 0.5g or less fat per serving.
Fatty acid	A long carbon chain carboxylic acid. Fatty acids vary in length and in the number and location of double bonds; three fatty acids linked to a glycerol molecule form fat.
Feed	Edible materails which are consumed by animals and contribute energy and/or nutrients to the animals' diet."
Feints	Feints are the third fraction of the distillation cycle derived from the distillation of low wines in a pot still. This scotch term is also used to describe the undesirable constituents of the wash that are removed during the distillation of grain whiskey in a continuous patent still (coffey). These are mostly aldehydes and fusel oils.
Fermentable sugars	Fermentable sugars like glucose maltose, and maltotriose, can be fermented by distiller's yeast.
Fermentation kinetics	Fermentation kinetics is concerned with the rate of cell synthesis and/or fermentation product formation and the effect of environment on these rates.
Fermentation	A catabolic process that makes a limited amount of ATP from glucose without an electron transport chain and that produces a characteristic end product, such as ethyl alcohol or lactic acid.
Fermented fruit beverage	It is a fruit juice, which has undergone alcoholic fermentation by yeast. The product contains varying amounts of alcohol.
Ferritin	A protein compounds that serves as the storage form of iron in the blood and tissues.

Term	Terminology
Fertility	It refers to ability of plant not able to set and mature fruits but develop a viable seed.
Fetal alcohol syndrome	A group of physical and mental abnormalities in the infant that result from the mother consuming alcohol during pregnancy.
Fiber	A lignified cell type that reinforces the xylem of angiosperms and functions in mechanical support; a slender, tapered sclerenchyma cell that usually occurs in bundles.
Fibril theory	When sugar is added to the pectin solution, it destabilizes the pectin-water equilibrium and the pectin conglomerates forming a network of fibrils through the jelly the network of fibril holds the sugar solution in the inter-fibrillar spaces.
Filtration	Filtration is defined as that unti operation in which the insoluble solid component of a solid liquid suspension separated through membrane.
Fire point	The temperature at which a substance, which ignited burns freely when the ignition agent is withdrawn.
Firming Agent	substance added to precipitate residual pectin, thus strenthening the supporting tissue and preventing its collapse during processing.
First class fruits	Fresh fruits which do not satisfy the minimum quality requirement of the extra class and are superior to lower class.
Fish gills	Respiration organ in fish.
Fixed oils, Pigments, Antioxidants	The non-volatile fractions of spices and plant materials obtained through solvent extraction is known as fixed oils, pigments and antioxidants.
Flaking of soybean	Heat treatment prior to extraction of oil, increases efficiency of oil extraction.
Flash Pasteurization	Fruit juice is heated for only a short time at a temperature higher (5.5°C) than the pasteurization temperature of the Juice (for a minute).

Term	Terminology
Flash point	The temperature under standardized conditions at which a liquid begins to evolve inflammable vapours.
Flavor material	Through chemical processes, chemically identical to the substance present in a natural product and intended for human consumption either processed or not; e.g., citral obtained by chemical synthesis or from oil of lemongrass through a bisulfate addition compound.
Flavoring Agent or Adjuvant	Substance added to impart or help impart a taste or aroma in food.
Flavour adjunct	A substance used in or with a flavour but not essentially a part of it. These include solvents, antioxidants, enzymes, adjusting agents, emulsifiers, and acidulants.
Flavour enhancer	A substance added to supplement, modify, or enhance the original taste and/or aroma of a food without imparting a characteristic taste or odor of its own.
Flavour extract	The dilute extract of flavour prepared from spices and aromatic plant parts usually in alcohol, fortified with WONF, if needed.
Flavour or Flavourant	A substance, added to food, whose significant function is to affect odour, imparting a characteristic flavor to that food.
Flavour reversion	This problem is unique to soybean oil and other linolenate containing oils. The off flavour has been described as beany and grassy and usually develops at low peroxide values.
Flavour	The sensation produced by a material taken into the mouth, perceived principally by the sense of taste and smell, but also by the common chemical sense produced by pain, tactile, and temperature receptors in the mouth.
Flavouring ingredients	Any single chemical entity or natural mixture added to food, drugs or other products taken in the mouth, the clearly predominant purpose and effect of which is to provide all or part of the particular flavor of the final product.

Term	Terminology
Flavours	Those mixtures of ingredients whose exact composition is usually known only to their suppliers, sold in bulk to food and beverages manufactures. They are to be labeled as flavours per CFR 21 part 101 and may contain adjuncts that are nonflavour ingredients.
Flexible packaging	Bags, envelops, pouches, sachets, wraps etc, made of easily yielding materials such as film, foil or paper sheeting which, when sealed, acquires pliable shape.
Flipper	Can appears normal, but when struck against a tabletop one or both ends become convex and springs or flips out, but can be pushed back to normal condition by little pressure such can is termed as flipper.
Flocculation	It refers to random aggregation reactions in the absence of denaturation.
Flour bleaching agent	The chemical substances used to oxidize carotenoid pigments to make maida creamy white.
Flour Treating Agent	Substances, added' to milled flour, at the mill to improve its color and/or baking qualities, inducing bleaching and maturing agents.
Flow point	The temperature at which upon heating the solidified fat becomes soft and flows downward through an orifice.
Fluorescence	Fluorescent compounds may develop from the interaction of carbonyl compounds (produced by lipid oxidation) and certain cellular constituents possessing free amino groups.
Foam spray drying	In foam spray drying a compressed gas is injected into the fluid prior to the spray nozzle. The injected air causes the solids associated with a droplet to agglomerate in a manner, which increases its buoyancy as compared with conventional spray drying.
Foam stability	The ability of foaming agents (flour or protein) to retain the foam volume upon standing.
Foam	Foam is dispersion of gas bubbles in liquid phase or semisolid phase.

Term	Terminology
Foaming capacity	The volume (ml) of foam formed when a known quantity of flour or protein is mixed with distilled water and stirred for 5 min.
Fold	Strength of concentrated flavouring materials. The concentration is expressed as a multiple of a standard, e.g., citrus oil is compared to cold pressed oil. In the case of vanilla, folded flavours are compared to a standard extract with minimum bean content.
Folding carton	Folding carton made from sheets of paperboard which have been cut and scored for bending into desired shapes, they are delivered in a collapsed state for erection at the packaging point.
Food additive	Food additives substances are incorporated in to foods for improve functional purpose, it can be found naturally in some food they are used in processed foods or these substances are intentionally added to food to prevent microbial spoilage.
Food chain	A feeding series in an ecosystem.
Food Chemistry	Food chemistry is major aspect of food science, which deals with the composition and properties of food and chemical changes it undergoes during handling, processing and storage.
Food intolerance	An adverse reaction to food that does not involve an immune response.
Food preservation	Science which deals with the methods of preservation of food material from decay or spoilage and allow it to be store in good conditions for future use.
Food retailing	Food retailing is a sale of food items in small lots to the consumers.
Food Science and Technology	It is a science deals with food and their technology.
Food Science	The Science deals with the food, which is edible by living matter.

Term	Terminology
Food sensitivity	A mild reaction to a substance in a food that might be expressed as slight itching or redness of the skin.
Food web	The elaborate, interconnected feeding relationship in an ecosystem.
Food	Any edible material, which is consumed by the individual and his/her family.
	Anything either solid or liquid which when consumed gets digested, provides energy, regulates growth, protects body and thus sustains life or An article that provides taste, aroma or nutritive value. Food and Drug Administrative (FDA) considers food as 'generally recognized as safe' (GRAS)."
	Whatever we eat and that nourishes the body is called food.
Food-borne illness	Sickness caused by ingestion of foods containing toxic substances produced by micro-organisms.
Foods for body building	Foods rich in protein, minerals, vitamins and water.
Foods for energy	Foods rich in carbohydrates, fats and protein.
Foods for regulating body process	Foods rich in minerals, vitamins, water and fiber.
Foots	The saponification fatty acids in crude oil which settle down to bottom on agitation of oil with alkali.
Fore milk	The first breast milk delivered in the nursing session.
Foreshots	Foreshots is the first fraction of the scotch distillation cycle derived from the distillation of low wines in a pot still.
Forgetting	The inability to recall a particular piece of information accurately.
Formulation Aid	Substance used to promote or to produce a desired physical state or texture in food. Including carriers, binders, fillers, plasticizers, film-formers, and tableting aids, etc.

Term	Terminology
Fortification	Addition of nutrients in amounts significant enough to render the food a good to superior source of the added nutrients. This may include addition of nutrients not normally associated with the food or addition to levels above that present in the unprocessed food.
Fortified	A term generally meaning that vitamins, minerals or both have been added to a food product in excess of what was originally found in the product.
Fracturability	The ease with which the material will break.
Free fatty acids	The fatty acids in the oil that are not esterified with glycerol and those liberated from triglycerides when subjected to hydrolytic rancidity.
Free radical	Short lived form of compounds that exist with an unpaired electron in their outer electron shell. This causes it to have an electron-seeking nature, which can be very destructive to electron-dense areas of a cell, such as DNA and cell membranes.
Free water	The water not bound to the components in a food. This is available for microbial use.
Freeze drying	In this process food is first frozen at -18 C on trays in the lower chamber of freeze drier and the frozen material dried under high vacuum in the upper chamber. Direct sublimation of ice taken place without passing through the intermediate liquid stage.
Freezer burn	Discoloration due to surface dehydration.
Freezing or Cooling Agent	A fuel is a substance (mostly carbon and hydrogen) which on burning with oxygen produces large amount of heat.
Freezing point	The freezing point of a material is the temperature at which it changes from a liquid to a solid. A liquid freezes when its vapour pressure is equal to the vapour pressure of its solid. The freezing point of water is 0°C.
Fructose	A monosaccharide with six carbons that form a five-membered ring with oxygen in the ring; found in fruits and honey.

Term	Terminology
Fruit cordial	This is fruit squash from which all suspended material is completely eliminated and is perfectly clear, e.g., limejuice cordial.
Fruit drink	Liquefying the whole fruit makes this and at least 10% of the volume of undiluted drink must be whole fruit. It may be diluted before being served.
Fruit juice concentrate	This is fruit juice, which has been concentrated by the removal of water either by heat or by freezing.
Fruit juice	This is a natural juice pressed out of a fruit, and is unaltered in its composition during preparation and preservation.
Fruit punches	Mixing the desired fruit juices at the time when it is served makes these.
Fruit squash	This consists essentially of strained juice containing moderate quantities of fruit pulp to which sugar is added for sweetening, e.g., orange squash, lemon squash, mango squash, etc.
Fruit	A mature ovary of a flower that protects dormant seeds and aids in their dispersal.
	The development of ovary and jaunt tissues following blossoming.
Fumigant	Volatile substance used for controlling insects or pests.
Fumigations	The process of giving gaseous flow controlling insect's in storage called as fumigations.
Functional fibers	Functional fibers consisting of isolated, non-digestible carbohydrates which have beneficial physiological effects in humans.
Functional Food	This is similar to conventional food and is consumed as a part of normal diet but is known/ proven physiological benefits and capacity to reduce risk of chronic dieases beyond basic nutritional functions.
	Foods with a specific additional benefit that goes beyond the nutritional benefits of the nutrients they contain. Or Food that encompasses potentially healthful products, including any modified food or food ingredient that may provide a health benefit beyond that of the nutritional nutrients it contains.

Term	Terminology
Fungal rotting	The final stage of spoilage by molds.
Fusel oil	Fusel oil is an inclusive term for heavier, pungent tasting alcohols produced during fermentation. Fusel oils are composed of mixture of n-propyl, isobutyl, and isoamyl alcohols.
Fusible plug	A device used to put off fire in the furnace of boiler.
F-Value	It is the time in minute required to destroy the organism in specified medium at 230°F or 121°C.
Galactosemia	A disease characterized by the buildup of the monosaccharide galactose in the bloodstream resulting from the inability of the liver to metabolize it. If present at birth and left untreated, this disease causes severe growth and mental retardation in the infant.
Gamma rays	Forms of electromagnetic radiations of extremely short wavelength, similar to X-rays emitted by radioactive isotopes such as cobalt-60 that are used as source of energy in food irradiation in medical treatment.
Gelatinization	The tendency of starch molecules to swell and form gel when starch solution is heated.
Gelation	Gelation is the irreversible change in viscosity of egg yolk after thawing.
	When denaturation molecules aggregate to form an ordered protein network the process is referred to as gelation.
Gene cloning	Transfer of sequence of DNA (gene) from one organism to other organism.
Gene stability	Major of resistance to change with time of the sequence of gene within DNA molecule or of nucleotide sequence with gene.
Gene	Particular segment of DNA molecule or it is part of chromosome and responsible for particular character.
Generally recognized as safe (GRAS)	A group of food additives that in 1958 were considered safe, therefore allowing manufactures to use them thereafter when needed in food products. The FDA bears responsibility for providing they are not safe and can be remove unsafe products.

Term	Terminology
Generation time	The time required by micro-organism to double its population.
Genetic Engineering	This refers to the purposeful manipulation of genetic material to alter the characteristics of an organism in a desired way.
Genetic map	A diagram showing the relative sequence and position of specific genes along a chromosome or Geentic map is a description of how genes lie along of chromosome.
Genetics	The branch of biology concerned with study of heredity and variation.
Genotype	Genotype means complete genetic makeup of an individual.
Germination	It is the process in which soaked grains are allowed to sprout and develop vegetative growth or roots.
Gills	Sexual spore, i.e., basidiospores are formed, borne on the underside of fruiting bodies, mostly seen in Agaricus sp. of mushroom.
Glazed fruits/vegetables	Covering of candied fruits/vegetables with a thin transparent coating of sugar, which imparts them glossy appearance, is known as glazing fruits/vegetables.
Glucagon	A hormone made by the pancreas that stimulates the breakdown of glycogen in the liver into glucose; this raises the blood glucose level. Glucagon also performs other functions.
Gluconeogesis	The production of new glucose molecules by metabolic pathways in the cell. The source of the carbon atoms for these new glucose molecules is usually amino acids.
Glucose syrup	Sweet syrup prepared by hydrolyzing corn starch which mainly consists of glucose or Aqueous viscous solution of maltose, dextrose and other sugars.
Glucose	A six-carbon atom carbohydrate found in blood and in table sugar bound to fructose; also known as dextrose, it is one of the simple sugars.

Term	Terminology
Glycemic index	A ratio used to measure the relative ability of a carbohydrate to raise blood glucose levels as opposed to the ability of white bread (or glucose) to raise blood glucose levels.
Glycerol	An alcohol containing three hydroxyl groups (–OH); used to help from triglyceride molecules.
Glycogen	An extensively branched glucose storage polysaccharide found in the liver and muscle of animals; the animal equivalent of starch.
Glycolysis	The splitting of glucose into pyruvate. Glycolysis is the one metabolic pathway that occurs in all living cells, serving as the starting point for fermentation or aerobic respiration.
Goiter	An enlargement of the thyroid gland often caused by a lack of iodine in the diet.
Goitregens	Substances in food that interferes with thyroid hormone metabolism and so may cause goiter if consumed in large amounts.
Grading	Grading is separation of harvested material in desired groups according to the market perference. Sorting on the basis of quality criteria.
Grain whiskey	Grain whiskey as an alcoholic distillate from fermented wort derived from malted and unmalted barley and corn, in varying proportions, and distilled in a continuous patent still (Coffey).
Growth	The progressive development or increase in size of a living thing, such as the growth of a child.
Gumminess	The energy required to break down a semi-solid food ready for swallowing.
Gustation	A taste sense, the receptors of which lie in the mucous membrane covering the tongue, and the stimuli for which consist of certain soluble chemicals, e.g., salts, acid, and sugar.
Haemoglobin	The oxygen-carrying pigment of red blood cells
Halcoarbons	These are obtained after replacing one or more hydrogen atoms in hydrocarbon with halogens.

Term	Terminology
HANES	Health and Examination Surveys: A survey conducted between 1971-72 on the United States Population aged 1 to 74 years to evaluate their nutritional status. The sample size of the population was about 20,000.
Hard to cook legume	Rate of water absorption and cooking is low, retards cooking.
Hard water	Water containing chlorides and sulphates of calcium and magnesium.
Hardness	The force required to compress the material by a given amount.
Harris-Benedict equation	An equation that predicts resting metabolic rate based on a person's weight, height, and age.
Harvest index	Harvest index is a measurement that can be used to determine whether a particulars commodity is mature or not.
Harvesting	Removal of Economical plant part at its maturity from the plant is called as harvesting or Removal and picking up of economically important part of crop plant for various purposes.
Haugh unit	Measurement of height of thick white in relation to the weight of egg.
Heads	Heads is distillate containing a high percentage of low boiling components such as aldehydes.
Health food	Health food refers to specific foods claimed to be especially beneficial to health for a long time.
Health	A state of complete physical, mental and social well-being and not merely the absence of disease or infirmity (WHO, 1948).
Heartburn	A pain emanating from the esophagus, caused by stomach acid backing up into the esophagus and irritating the esophageal tissue.
Heat pump	The machine which performs heating process.
Heat transfer coefficient	Heat transfer coefficients for drop wise condensation of vapour. The heat transfer coefficients for drop wise condensation are very high in the range of 10,000 to 70,000 Btu/hr. (sq.ft) (°F).

Term	Terminology
Heat transfer	The transfer of heat is the principal unit operation in many food processes.
Heat	Heat is energy transferred from one system to another solely by reason of a temperature.
Hehner number	The percentage of water and fatty acids in 1g of fat.
Hemoglobin	The iron-containing protein in the red blood cell that carries oxygen to the cells and carbon dioxide away from the cells. It is also responsible for the red colour of blood.
Hemolysis	It is the process where by the red cell bursts, releasing the haemoglobin from it
Hemorrhage	It is a condition in which blood escapes from the vessels. Some time it also called as bleeding.
Herd immunity	Resistance of a group to a pathogen as a result of immunity of a large proportion of the group to that pathogen.
Heredity	Transmission of character from parent to offspring is called heredity.
Heritability	Heritability is defined as the portion of phenotypic variation that is due to additive effect of genes.
Heterologous probes	The probes obtained from one species and used for another species.
Heterotrophic nutrition	A mode of obtaining organic food molecules by eating other organisms or their by-products.
High wines	High wines is an all-inclusive term for beverage spirit distillates that have undergone complete distillation.
High-density-lipoprotein (HDL)	The lipoprotein synthesized by the liver and intestine that picks up cholesterol from drying cells and other sources and transfers it to the other lipoproteins in the bloodstream. A low HDL level increases the risk for heart disease.
Hind milk	The milk secreted at the end of a nursing session; it is higher in fat than fore milk.

Term	Terminology
Hintons theory	It is based on the assumption that pectins are complex mistures of variable composition. Gelating of pectin is a type of coagulation in which the coagulated particles form continuous networks.
HIV (Human Immunodeficiency Virus)	The infectious agent that causes AIDS; HIV is an RNA retrovirus.
HLB value	Hydrophile-lipophile balance is ratio of weight of percentage of hydrophilic and hydrophobic groups in an emulsifier.
HMG-CoA reductase	An enzyme in the cytosol that catalyzes the conversion of hydroxymethylglutaryl-CoA (HMG-CoA) to form mevalonate. The action of HMG-CoA reductase is the committed step in cholesterol biosynthesis.
Holding cost	The cost associated with holding inventories in a stock is known as holding cost.
Hops	Hops are dried female flowers of hope plants which are grow in Ciregon and Washington.
Hormone	A compound secreted into the bloodstream that acts to control the function of target organ cells. Hormones can be either protein like or fatlike, such as insulin or estrogen.
Hue	Hue is the aspect of colour that we describe by words such as green, blue, yellow and red.
Humectant	Hygroscopic substances incorporated in food to promote retention of moisture. Includes moisture retention agents and antidusting agents.
Hydrogenation of oil	It is the process in which liquid fat or oil is converted into solid or semi-solid fat by addition of hydrogen.
Hydrogenation	The chemical addition of hydrogen to a material or The addition of hydrogen atoms to the double bonds of polyunsaturated and monounsaturated fatty acids to reduce the extent of unsaturation. This process turns liquid vegetable oils into solid fats.

Term	Terminology
Hydrophilic- Lipophilic balance (HLB)	It is a ratio between hydrophilic emulsifier and lipophic emulsifier. As rule emulsifiers with HLB values in the range 3-6 promote W/O emulsions; values between 8 and 18 promote O/W emulsions.
Hydrophilic	Attracts water (water loving).
Hydrophobic	Repels water (water fearing).
Hyperglycemia	High blood glucose levels, above 140 milligrams per 100 milliliters of blood.
Hyperosmia	Unusually keen olfactory sensitivity.
Hypertension	A condition in which blood pressure remains persistently elevated, especially when the heart is between beats.
Hyphae	The mold thallus consists of a mass of branching interwined filaments called hypahe.
Hypobaric storage	The storage in which pressure in the storage atmosphere is less than atmospheric pressure.
Hypogeusia	Diminished sense of taste.
Hypoglycemia	Low blood glucose levels, below 4 to 50 milligrams per 100 milliliters of blood.
Hyposmia	Diminished sense of smell.
Hypothesis	A proposition or assertion about the possible relationship between variables.
Ice cream	Ice-cream is the frozen product obtained from cow or buffalo milk or a combination thereof or from cream, other milk products, with or without the addition of cane sugar, fruits, juices, nuts, edible flavours and permitted colours.
Imbibed water	May not be different from water held as hydrate some substances pickup water and swell, when they come in contact with water.
Immobilized enzymes	An immobilized enzyme or cell is chemically or physically restricted in movement so that it can be physically reclaimed from the reaction medium.
Immune system	A complex of organs and cells that function to protect the body against disease agents such as viruses, bacteria and pollutants.

Term	Terminology
Impact resistance	Relative susceptibility of material to fracture by stress at high speeds.
Imperfect flower	Species that have only stamens or stigma within a flower.
Incidental food additives	Additives that gain access to food products indirectly from environmental contamination of food ingredients or during the manufacturing of food process.
Incineration	The process fo burning wastes under controlled condition.
Income terms of trade	This is obtained by dividing the value of export (value index = quantity index and price index) divided by index of import prieces.
Incomplete proteins	Food protein that lacks ample amount of one or more of the essential amino acids needed to support human protein needs.
Index of nutrition quality	It is percentage of nutrient need provides by food divided by percentage of calorie need provided by from food.
Infusion mashing	Infusion mashing is the process of simultaeously cooking and converting small grains (rye, barley, and wheat).
Inherent toxicants	These are metabolites produced via biosynthesis of food under normal growth conditions and by organisms.
Inland fishery	Process of raising and harvesting fresh water fish, rivers, lakes are the places of habitat of fresh water fish, e.g., rohw, katla and mrigal etc.
Insipid	Tasteless, fat, vapid.
Insoluble fiber	Fibers that, for the most part, do not dissolve in water nor are digested by bacteria in the large intestine. These include cellulose, some hemicelluloses and lignin.
Instant dhal	Less time required for cooking as compared to ordinary dhal/ready to eat dhal.
Instrumentation	The technology, of using instruments and control the physical chemical properties of materials is called instrumentation.

Term	Terminology
Insulin	A hormone produced by the beta cells of the pancreas. Insulin increases the synthesis of glycogen in the liver and the movement of glucose from the bloodstream into muscle and adipose cells.
Integral protein	A protein of biological membranes that penetrates into or spans the membrane.
Intelligence Quotient (IQ)	The relationship between mental age and chronological age, expressed as MA/CA x 100.
Intelligence	The capacity to understand the world and the resourcefulness to cope with its problems.
Intermediate density lipoprotein (ILD)	The product formed after a very low-density lipoprotein (VLDL) has most of its triacylglyceride removed.
International unit (IU)	A crude measure of vitamin activity, often based on the growth rate of animals. Today these units have been replaced by more precise microgram quantities.
Internet	It is worldwide nework of computers.
Intoxication	The state in which an excess of a sedative poisons the body, leading to uncoordinated actions, slowed reaction times and slurred speech.
Inventory	It is aggregate investment item of tangible personal property that are sale production for sale and production of goods available for sale after consumption for production.
Invert sugar	The hydrolyzed mixture of sucrose, containing glucose and fructose is known as invert sugar.
Iodine value	The number of iodine absorbed by 100 g of oil. It is a measure of the unsaturation of fat.
	This test is a measure of the unsaturated linkages in a fat and is expressed in terms of percentage of iodine absorbed.
Iron	A metallic element occurring in the haemoglobin of red blood cells, stored in tissues in the form of ferritin and is an essential part of important respiratory enzymes

Term	Terminology
Irradiation	Employ radiant energies in the form of X-rays that produce their effects upon being absorbed within the food.
Isoelectric point pH	It is defined as the pH at which a solute (compound) has no net electric charge or carry equal positive and negative charge (i.e., zwitterions).
Isolate	A relatively pure chemical produced from natural raw materials by physical means, e.g., distillation, extraction, crystallization, etc. and therefore natural; or by chemical means, i.e., via hydrolysis, bisulfate addition products, and regeneration, etc. and therefore artificial by 1993 U.S. labeling regulations.
Isolated system	A system which is completed uninfluenced by surrounding.
Isomer	Different chemical structures for compounds that share the same chemical formula.
Isotope	An alternate form of a chemical element. It differs from other atoms of the same element in the number of neutrons in its nucleus.
Isozymes/Isoenzymes	Multiple forms of the same enzyme arising from genetically determined differences in primary amino acid sequences.
ITR	It is equivalent to the of heat transfer needed to produce 1 ton of ice at 32 °F from water at 32°F in 1 day.
Jam	It is a product prepared by boiling fruit pulp with sufficient quantity of sugar to reasonably thick consistency, firm with enough to hold fruit tissues in position. It should contain not less than 68.5% T.S.S.
Jaundice	A yellow staining of the skin and sclera resulting from a buildup of bile pigments in the bloodstream. Liver or gallbladder disease is often the cause.
Jelly	It is a product prepared by mixing strained extract of fruit with sugar and boiling the mixture to a stage at which it will set a clear gel.

Term	Terminology
Juice	Juice is the liquid that is naturally contained in fruit and vegetables. It can also refer to liquids that are flavored with these or other biological food sources such as meat an seafood. It is commonly consumed as a beverage or used as an ingredient or flavoring in foods.
	Juice is prepared by mechanically squeezing or macerating fruit or vegetable flesh without the application of heat or solvents. For example, orange juice is the liquid extract of the fruit of the orange tree, and tomato juice is the liquid that results from pressing the fruit of the tomato plant. Juice may be prepared in the home fresh fruit and vegetables using a variety of hand or electric juicers. Juice is one of the most popular drinks to go with breakfast in the morning.
Junk DNA	The genetic material of animal and plants consists of mainly two types coding and non-coding. The non coding is also known as no functional or junk DNA.
Kernel Hardness	Kernel hardness measurements are mainly useful in differentiating between soft or hard grains in plant breeding programme. Extensively hard reflects in to increase in power for grinding while extensively softness reflects into bolting and increases the requirements for sieving space.
Ketone bodies	Products of acetyl-CoA (Fat) metabolism containing three to four carbon atoms; acetoacetic acid, beta-hydroxybutric acid, and acetone. These contain a ketone group, hence, the name.
Ketosis	The condition of having high levels of ketones in the bloodstream.
Key column	The optimum column, the one with largest negative index number.
Kirchhoff's law radiation	The ratio of total emissive power to absorptivity is constant. For all substance which are in thermal equilibrium the surroundings.

Term	Terminology
Kreis test	This is one of the first tests used commercially to evaluate oxidation of fats. It measure a red colour believed to result from reaction of epihydrin aldehyde or other oxidation products with phloroglucinol.
	A test in which the oxidized fats react with phloroglucinol in acid solution to give a red colour.
Kwashiorkor	A disease seen primarily in young children who have an existing disease and who consume a marginal amount of calories and considerably insufficient protein in the face of high needs. The child will suffer from infections and exhibit edema, poor growth, weakness, and an increased susceptibility to further illness.
	Means a sickness in a child develops when another baby is born.
Lamination	Combining two or more films into single film by means of solvent or heat to compliment all the properties in a single film is called lamination.
Lard	It is an animal fat from hogs.
LDL (low density lipoprotein) cholesterol	The fraction of cholesterol carried in the blood as a part of low-density lipoproteins.
Lean body mass	The part of the human body that is free of all but essential body fat. About 2% of body weight as fat is essential to retain. The rest of the fat in the body represents storage and so is not part of lean body mass. Lean body mass includes muscle, bone, organs, connective issue, skin and other body parts.
Lean fat	Meat, poultry, seafood, and game meats containing <10g of fat per serving.
Leathery crust	Crust of bread is crisp and breaks easily when pulled but crust becomes tough and exhibits the quality of leatheriness when pulled.
Leaving Agent	Substance used to produced or stimulate production of carbon dioxide in baked goods in order to impart a light texture, including yeast, yeast foods, and calcium salts.
Lecithin	A group of phospholipids containing two fatty acids, a phosphate group, and a choline molecule.

Term	Terminology
Lectins	Antinutritional factor present in plant foods some time it is called hemaglutinins.
Legumes	Legumes are the dicotyledonous seeds of the pod bearing plants belonging to the family *Leguminosae*.
Liarrage	The place where animals are brought 24 hrs before slaughtering.
Life span of spore	The life span of spore is defined as the length of time for which spores exposed to given temperature, remain viable.
Light fat	An altered product containing one-third fewer calories or 50% of the fat in the reference food.
Lignin	An insoluble fiber made up of a multiringed alcohol (nocarbohydrate) structure.
Limit dextrin	Limit dextrin are oligosaccharides containing one or more 1,6-a-linkages.
Limiting amino acid	The essential amino acid in the lowest concentration in a food in comparison with the body's need.
Line	A group of individual having common parents or ancestors.
Linkage map	A map which depicts the order of genetic marker and the relative distances between them as measured in terms of recombination frequencies between the markers.
Lipogenesis	The building of fatty acids using derivatives of acetyl-CoA molecules.
Lipogenic	Means creating lipid. The liver is the major lipogenic organ in the human body.
Lipolysis	Hydrolysis of ester bonds in lipids may occur by enzyme action or by heat and moisture, resulting in the liberation of free fatty acids. The release of short chain fatty acids by hydrolysis is responsible for the development of a rancid flavour (hydrolytic rancidity) in raw milk. Lipolysis is a major reaction occurring during deep fat frying due to the large amounts of water introduced from the food and the relatively high temperature at which the oil is maintained.

Term	Terminology
	Hydrolysis of lipids to produce glycerol and free fatty acids.
	The breakdown of lipids.
Lipoprotein	A compound found in the bloodstream containing a core of lipids with a shell of protein, phospholipid and cholesterol.
Locus	Point occupied by a gene on chromosome.
Low calorie	A product with<40 cal per serving.
Low fat	A serving containing no> 40 cal, and 3g or less fat.
Low wines	Low wines is the term for the initial product obtained by separating (in a pot or Coffey still) the beverage spirits and congeners from the wash. Low wines are subjected to at least one more pot still distillation to attain a greater degree of refinement in the malt whiskey.
Lozenges	These are the sugar dough which has been flavoured, cut to shape and subsequently dried to remove the most of the added water.
Lubricant or Release Agent	Substance added to food contact surfaces to prevent ingredients and finished products from sticking to them (direct additives). Includes releasar agents, lubricants, surface lubricants, waxes and anti-blocking agents (indirect additives).
Macrosmatic	Abnormally keen olfactory sense.
Malnutrition	Impairment of health resulting from a deficiency or imbalance of nutrients. It can be under nutrition or over nutrition.
	The chronic deficiency of one or more nutrient in daily diet.
Malt whiskey	Malt whiskey is an alcoholic distillate made from fermented wort derived from malted barley only and distilled in pot stills. It is the second fraction (heart of the run) of the distillation process.
Malting or Fermenting Aid	Substance used to control the rate of nature of malting or fermenting proccess including microbial nutrients and suppressants and excluding acids and alkalis.

Term	Terminology
Malting	It is processing of grain where grains are steeped, germinated and kilned to increase maltose content of grain.
	Malting is a controlled germination process, which activates the enzymes of the resting grain, resulting in conversion of cereal starch to fermentable sugars, partial hydrolysis of cereal proteins and other macromolecules.
Management	It is the administrative of business concerns of public undertaking or it is a process b y which a co-operative group directs action towards common goals.
Manipulated variable	The variable that affects the value of the measured or controlled variable is called the manipulated variable.
Marasmus	Results from a person consuming insufficient protein and kcalories; usually seen in infancy. It is the equivalent of protein-energy malnutrition in adults. The person will have or no fat stores and show muscle wasting.
Margarine	Margarine is made from vegetable oils or a mixture of vegetable and animal fat by hydrogenation. Margarine used as a substitute for butter.
	Mixtures of fats blended to provide desired properties for use as table spreads or for bakery use.
Market testing	It is the introduction of new products into regions selected for a variety of geographical marketing and company reasons.
Marketing channels	It is the chain of intermediaries through whom the various food grains pass from producer to consumers.
Marmalade	It is a fruit jelly in which slices of fruit or of the peel are suspended. The term marmalade is generally associated with product made from citrus fruit like oranges. Lime/Lemon shredded peel is included as the suspended materials.

Term	Terminology
Mashing	Mashing is the process to soubise as much portion as possible of malt and malt adjuncts and especially to hydrolyze starches, polysaccharides and proteins.
Masticatory Substance	Substance that is responsible for the long lasting and pliable property of chewing gum.
Maturity	Stage of development giving minimum acceptable quality to the ultimate consumer sufficient stage of development that after harvesting and post harvest handling.
Mayonnaise	An emulsion of oil-in-water that readily breaks if crystals begin to form in oil.
Meat	Flesh of the animal used as food.
	Meat may be defined as contractive tissue from all animals used for food.
Megaloblast	A large, immature red blood cell that results from an inability for cell division during red blood cell development.
Melanoids	The brown to black amorphous, unsaturated heterogeneous polymers are called melanoids.
Melting point	The temperature at which a fat becomes completely clears.
Memory	The ability to store information so that it can be used at a later time.
Mental age	A measure of an individual's performance on an intelligence test expressed in terms of months and years.
Merosmia	A condition analogous to colour blindness, in which certain odors are not perceived.
Metabolic engineering	The science which study the effect of metabolic energy on overall metabolism, their enhancer, the release of energy.
Micro-filtration	Micro filtration involves the removal of very fine particles of the order of 1 mm or less and generally directed at removing microorganisms from liquid foods.

Term	Terminology
Micro-aerophillic	Microbe requiring small amount of air.
Microbiology	The discipline of science dealing with the laws of life and development of organisms particularly microsscopics forms of life, which are usually invisible to the naked eye, is called microbiology.
Microsmatic	Having a poorly developed sense of smell.
Mid day meal programme	It is a programme run under Govt. schools for development of nutritional status and attracts children from rural sector towards schools.
Migration	Establishment of plant or animal species in a new area.
Milk fats	Milk fats mostly contain palmitic, oleic and stearic fatty acids and its fat is unique among animal fats in that it contains C4 to C12 chain fatty acids.
Milk	Milk is normal secretion of mammary gland is defined as the lacteal secretion. Practically free from colostrums, obtained by complete milking of one or more than one cows which contains not less than 8% MSNF and not less than 3 - 35.% milk fat.
	Milk is the whole, fresh, clean lacteal secretion obtained by the complete milking of one or more healthy milch animals, excluding that obtained within 15 days before or 5 days after calving.
Mill feed	Mixture of bran, germ and shorts.
Millard reaction	The browning reactions occur in the presence of an amino bearing compound, usually a protein, a reducing sugars and some water.
	It changes colour, lavor, odour and texture of the product. These types of reaction occur by different four ways.
	1. Sugars + nitrogenous compounds + heat
	2. Sugar + organic acids + heat
	3. Nitrogenous compounds + organic acids + heat
	4. Organic acids themselves + heat

Term	Terminology
Milling loss	Loss of moisture by evaporation during milling.
Milling	Process of separation of bran and germ from endosperm and reduction of endosperm into fine flour.
Minimum support price	This is the price fixed by the Government to protect the producer. Farmers against excessive fall in the price during bumper production year.
Miscella	After extraction mixture of oil, hexane, other material is called miscella.
Miso	Fermented food product prepared from soybean and rice using microorganisms for fermentation process.
Mixing	Two or more components are interspersed in space with one another by means of flow.
Molar extinction coefficient	In the specific absorption coefficient for a concentration of one mole and path length 1 cm.
Molecular biology	The study of biology at molecular level is molecular biology.
Monod equation	$\mu = \mu\ max - S/ks + S$
Motivation	The desires, needs and interests that arouse an organism and direct it toward a specific goal.
Mould	Fungi that have microscopic size and filamentous multicellular organization.
	Moulds are multicellular, filamentous fungi belonging to the division Thallophyta but are devoid of chlorophyll.
	Penicillium sp. Blue moulds
	Aspergillus sp. Black moulds
	Mucor sp. Gray moulds
mRNA	As template used by ribosome for the translation of the genetic information into amino acid sequence of proteins.
Mulching	It is system of orchard soil management in which orchard soil is covered with curry material to reduce moisture loss due to evaporation.

Term	Terminology
Multiple personality	A form of dissociative disorder in which large segments of the personality are split off from conscious awareness, so that the person seems to fluctuate between two or more distinct personalities.
Mutarotation	The change in specific optical rotation representing inters conversions of a and b forms of D glucose to an equilibrium mixture.
Mutation	Alterantion in the DNA structure that produces permanent change in the genetic information.
Mycotoxin	Toxin produced by molds, e.g., Aflatoxin.
Myosin	Myofibrillar thick filament protein.
NDF (neutral detergent fiber)	The representative of the fibrous cell wall constitutes, containing ignin, cellulose, hemi-celluloses and some fiber-bound proteins.
Nanotechnology	Nanotechnology deals with developing materials, devices, or other structure with at least one dimension sized from 1 to 100 nanometers.
Natural food	Natural food implies to foods that are minimally processed and do not contain manfuactured ingredients, but the lack of standards in some jurisdictions (*e.g.,* fresh cut vegetables).
Natural resource	Life on this eart depends upon a large number of things and services provided b y the nature.
Nature identical	A flavour ingredient obtained by synthesis, or isolated from natural products
NCHS	National Centre for Health Statistics, United States—NCHS collected and analyzed statistics on health status, human needs and resources. NCHS also published data in the form of tables on weight and height for children and adolescents upto 18 years old (sex separated)
NDP (net dietary protein) cal%	The percentage of total calories proved by the protein.
Nectar	Nectar is obtained by blending the thin pulp of the fruit with sugar and citric acid. The finished product has 15-20 Brix and a mild acid taste.

Term	Terminology
Nectar	Type of fruit beverage which contains at least 20% fruit juice/pulp 15% TSS and 0.3% acid.
Neurotoxin	Toxin affects central nervous system.
Neurotransmitter	A chemical substance involved in the transmission of neural impulses from one neuron to another. Neurotransmitters are released when an action potential reaches the end of the axon (axon terminal). Upon release, neurotransmitters cross the synapse and attach to receptors on either the dendrite or cell body of the adjacent neuron.
Neutralizatior. of oil	Process for removal of free fatty acids from oil.
Newtonian fluid	Fluid obeys Newton's viscosity law.
	For the fluids the ratio of shear stress to shear rate is constant and equal to the viscosity of the fluid. Such fluids are called Newtonian fluids.
Noise pollution	The unpleasant and unwanted sound.
Nonbreak oil	Edible oil, which does not deposit solid material when heated to at least 300°C.
Non-climacteric fruits	Fruits those are showing respiratory pattern slow drift down wards after detachment from parent plant.
Nonnutritive Sweetener	Substance having >2% of the caloric value of sucrose per equivalent unit of sweetener capacity.
Nosocomical infection	Hospital acquired infection.
Nougats	These are the confections prepared by using sugar, glucose syrup and whipping agents which increase the desired volume of air.
NPR	Weight gain in test protein group (g) + weight loss in non-protein group (g) ÷ Protein consumed (g).
NPSH	The NPSH is the amount by which the pressure at the suction point of pump (i.e., sum of velocity and pressure heads) is in excess of the vapour pressure of the liquid.
NPU	N intake-fecal N-urine N × 100 ÷ N intake + (metabolic N + endogenous N) × 100 ÷ N intake.

Term	Terminology
Nuclear magnetic resonance (NMR)	Brain-imaging technique in which the head is exposed to magnetic fields of varying strengths and a computer-assisted three-dimensional image of the brain is produced.
Nucleoside	A combination of nitrogen base and sugar is called nucleoside.
Nutraceuticals	Nutraceuticals are food supplements or a part/ constituent of a food with specific medical or health benefits, for prevention, treatment or cure of a disease of
	A nutraceutical can also be defined as "a diet supplement that provides a concentrated form of bioactive compounds, of food, in a non-food matrix to enhance health and prevent diseases."
	The meaning was modified by Health Canada which defines nutraceutical as: a product isolated or purified from foods, and generally sold in medicinal forms not usually associated with food and demonstrated to have a physiological benefit or provide protection against chronic disease.
	Nutricuticals are products and their derivatives that occur in Nature and are constituents of plants and animal, including humans. These constituents confer a health benefit above and beyond basic nutrition or basic fortification.
Nutrient Supplement	Substance necessary for the body's nutritional and metabolic process.
Nutrient	Any substance assimilated by organism that promotes growths, generally applied to nitrogen and phosphorus in waste water but also to other essential and trace element.
	The combination of processes by which a subject receives and utilizes materials (foods) necessary for the maintenance of body components.
	Nutrients are the constituents in food that has specific function in body.
Nutrition assessment	Includes the measurement and description of the nutritional status of an individual or population in relation to economic, socio-demographic and physiological variables.

Term	Terminology
Nutrition	It is an science of food and its interactions with an organism to promote and maintain health.
Nutritional status	The condition of the body of individual(s) resulting from intake, absorption and utilization of food or nutrients over a period of time.
Nutritionist	A person who advise about nutrition and/or works in the field of food and nutrition. In many states in the United States a person does not need formal training to use this title. Some states reserve this title for Registered Dietitians.
Nutritive Sweetener	Substance having>2% of sucrose per equivalent unit of sweetening capacity.
Nyctalopia	It is night blindness, failure or imperfection of vision at night or in a dim light.
Obesity	A condition characterized by excess body fat, usually defined as 20% above desirable weight.
	When the energy intake exceeds over expenditure, the excess is deposited as a fat over a period of time obesity occurs.
Objective evaluation	Objective evaluation involving instruments. These instruments may be categorized into two types, viz., imitative measures and non-imitative measurements.
Odor and odorant	That which is smelled. Odor may refer to the odorant or to the sensation resulting from the stimulation of olfactory receptors in the nasal cavity by gaseous material.
Odoriphore (Osmophore)	Odor-producing group.
Oilseed meal	Oil see cake obtained during oil extraction, rich in protein.
Okazaki fragment	The lagging strand is formed by joining many small DNA fragments.
Oleoresins	The extracts from herbs or spices or other plant materials. These are complex mixtures obtained by solvent extraction. It constitutes volatile and non volatile fractions. It is usually viscous liquids or semi-solids.

Term	Terminology
Olericulture	It deals with cultivation of vegetable crops.
Oligosaccharides	Carbohydrates containing three to ten mono-saccharide units.
Olsen's theory	Sugar acts as dehydrating agent distributing the equilibrium existing between water and pectin.
Omega-3 (w-3) fatty acid	A fatty acid with its first double bond first appearing at the third carbon atom from the methyl end ($-CH^3$).
Operon	A unit genetic expression consisting of one or more related genes and the operator and promoter sequences that regulate their transcription.
Opsonization	Promotion of phagocytosis by a specific antibody in combination with complement.
Optimum feasible solution	Any basic feasible solution which optimizes objective function of general linear programming problem is called as optimum feasible solution.
Organic food	Organic foods are foods that are produced using organic methods.
Osmics	The science of smell.
Osmosis	The movements of micro-organism in response to salt or sugar solution.
Osmyl	An odorant.
Osteoporosis	A bone disease that develops primarily after menopause in women and is characterized by a decrease in bone density.
Oven spring	It is phenomenon of increases in loaf volume of dough during first 15 minutes of baking.
	Sudden increasing volume of dough during first 10-12 minutes of baking due to increase rate of fermentation and expansion of gases.
Overweight	Weight in excess of the average for given sex, age and height in relation to the NCHS reference tables.
Oxidation	The loss of electron from a substance involved in a redox reaction.

Term	Terminology
Oxidizing or Reducing Agent	Substance, which chemically oxidizes or reduces another food ingredient. thereby producing a more sable product.
Oxirane test	This method is based on the addition of hydrogen halides to the oxirane group. Epoxide content is determined by titrating the sample with HBr in acetic acid, in the presence of crystal violet to a bluish green end point.
Oxygen absorption	The amount of oxygen absorbed by the sample as determined by the time to produce a specific pressure decline in a closed chamber or the time to absorb a pre-established quantity of oxygen under specific oxidizing conditions is taken as a measure of stability. This test has been particularly useful in studies of antioxidant activity.
Oxygen bomb method	The samples are dispersed in filter pulp in the glass liner of a sealed bomb at 50 psig oxygen pressures. The bomb is placed in boiling water and the time required to reduce the pressure to 2 psig is noted.
Packaging	It is an external means of preserving the food during storage, transportation, and marketing.
Papain	Papain is a proteolytic enzyme extracted from the latex of the skin of developing papaya fruits.
Parageusia	Gustatory disturbance resulting in erroneous identification of taste stimuli.
Parasite	A microbe which depends on living matter for its growth and survival.
Parboiling of legums	Partial boiling of legumes to reduces leaching losses.
Parboiling	It is a hydrothermal treatment given to paddy followed by drying before milling.
Parosmia	A disturbance to the sense of smell resulting in smelling the wrong odors, usually perceived as repulsive.
Partition chromatography	If the moving phase is liquid and the stationary phase is a liquid film on same support, the resulting chromatography is called partition chromatography.

Term	Terminology
Passive absorption	Absorption that uses no energy. It requires permeability for the substance through the wall of the intestine and a concentration gradient higher in the lumen of the intestine than in the absorptive cell.
Pasteurization	Pasteurization means making free the food from human pathogens and most vegetative microorganisms by heat treatment upto 100°C for shorter period.
	The process of heating the juice to 100°C or slightly below for a sufficient time to kill mciro-organisms which causes spoilage is called as paseurization.
Pathogencity	The ability of a parasite to infect or to cause the disease in host.
Pelagic fish	Fishes that found on ocean surface.
Pellagra	A disease characterized by inflammation of the skin, diarrhea and eventual mental incapacity resulting from the lack of the vitamin niacin in the diet.
	Disease associated deficiency of nicotinic acid, characterized by dermatitis, diarrhea, and dementia, usually seen in area where sorghum is staple food.
PEM	Protein energy malnutrition is defined as protein energy malnutrition occurring due to deficiency of protein and energy in the supplied food.
Pepsin	A protein-digesting enzyme produced by the stomach.
PER	The gain in body weight per gram of protein or Mean gain in body weight of rat (g) ÷ Protein consumed (g).
Percent fat free	A product must be low fat or fat-free, and the percentage must accurately reflect the amount of fat in 100g of the food.
Percentile	A value on a scale of one hundred that indicate the percent of a distribution that is equal to or below it.
Perception	The process, by which the organism selects, organizes and interprets sensations.

Term	Terminology
Pernicious anemia	The anemia that results from a lack of vitamin B-12 absorption. It is pernicious (deadly) because of the associated nerve degeneration that can result in eventual paralysis and death.
Peroxide value	It is expressed as milliequivalents of oxygen present in one kilogram of fat.
	Peroxides are the main initial products of autoxidation. They can be measured by techniques based on their ability to liberate iodine from potassium iodide or to oxidize ferrous to ferric ions. Their content is usually expressed in terms of milliequivalents of oxygen per kilogram of fat.
PGI2 (prostaglandin type I2)	A type of prostaglandin (hormone-like substance) derived from arachidonic acid.
pH Control Agent	Substance added to change or to maintain active acidity or basicity. Includes buffers, acids, alkalis and neutralizing agents.
Phenotype	Phenotype means the outward visible expression of the hereditary constitution of animal.
Phosphatides	A group of lipids which on hydrolysis give fatty acids, phosphoric acid and nitrogenous base.
Photo-oxidation	The oxidation of fatty acids by attack of singlet oxygen molecules produced from the usual triplet from by the action of light energy and sanitizers.
Photophosphorelation	The process of generating ATP from ADP and phosphate by means of a proton-motive force generated by the thylakoid membrane of the chloroplast during the light reactions of photosynthesis.
Photorespiration	The metabolic pathway that consumes oxygen, evolves carbon dioxide, generates no ATP, and decreases photosynthetic output; generally occurs on hot, dry, bright days, when stomata close and the oxygen concentration in the leaf exceeds that of carbon dioxide.
Phytate phosphorus	Phosphorus binds with phytic acid, no free phosphorus is available.

Term	Terminology
Phytic acid	Inositol hexaphosphoric acid.
Pickles	The process of preservation of food in common salt or in vinegar is called pickling and the product is called pickles. Spices and edible oil are added for taste.
Pitting of legumes	Scratching of seed coat for oil penetration in to seed and it is given as a premilling treatment to the legumes.
PKa	PKa is defined as the pH at which given ionic compound is half dissociated.
Plant growth regulators	These are enzymes or chemicals which stimulate the vegetative growth of plant.
Plant layout	The plan or act of planning an optimum arrangement of industrial facilities is called as plant layout.
Plasmids	Plasmids are small, self replicating extra chromosomal DNA.
Polenske value	The number of milliliters of 0.1 N alkalis required neutralizing the volatile water-insoluble fatty acids present in 5 g of fat.
Pollutant	A contaminant that adversely alter the physical, chemical or biological properties of environment.
Polymerization	Polymerization or aggregation reactions generally involve the formation of large complexes.
Polymorphism	Certain triglycerides exist in several different crystal systems each of which has a characteristics melting point, X-ray diffraction pattern and infrared spectrum. This phenomenon is known as polymorphism. Polymorphic forms are solid phases of the same chemical composition that differ among themselves in crystalline structure but yield identical liquid phases upon melting (a, b' and polymorphic structure).
Pork	Uncured meat of pig or swine.
Porosity	It is defined as the percentage of volume of grain bulk.

Term	Terminology
Post harvest technology	It is defined as techno economical activity applicable to all agricultural commodities or produced originated from farm, forest livestock and aqua culture for there conservation, handling, storage, marketing and value addition to make them useful as food, fed, fiber, fuel and industrial raw material.
Pour point	The lowest temperature at which a liquid will flow when a test container is inverted.
Poverty	A situation in which the level of living of an individual family or group is below the standard of living of the community either in terms of subsistence or in contrast to normal standards of income required for at least modest participation in community life.
Powder	When juice is converted into free flowing highly hydroscopic powder to which natural flavour in powder from is incorporated to compensate for any loss of flavour in concentration.
Power number	The ratio of external to internal forces per unit volume of liquid is defined as the power number.
Pre-cooling	Removal of field heat from the freshly harvested commodities (Fruits and vegetables) using cooling media.
Pre-packaging	Operation of packaging of fresh produce in thin plastic films.
Prebiotic	Prebiotic is 'a non-digestible food ingredient that beneficially affects the hosts by stimulating beneficial bacteria (e.g. dietary fiber).
	Prebiotics are a collective term for non-digestible but fermentable dietary carbohydrates that may selectively stimulate growth of certain bacterial group resident in the colon, such as biofidobacteria, lactobacilli and eubacteria, considered to be beneficial for human host.
Precipitation	It includes all aggregation reactions leading to a total or partial loss of solubility.
Precision	The coefficient of variability between duplicates may be used as estimate of precision.

Term	Terminology
Preservatives	Compounds that extend the shelf life of foods by inhibiting microbial growth or minimizing the destructive effect of oxygen and metals.
Preserves	It is a product made from properly matured fruit by cooking it whole or in the form of large pieces in heavy sugar syrup until it becomes tender and transparent. In this case cooking is continued until a concentration of T.S.S. is reached to 68.5° Brix.
Pressure drops	Pressure drops are the conditioned grains are subjects from high pressure to the very low pressure resultant product is puffed.
Primary metabolite	A metabolite produced by organism during the growth phase.
Primary packaging	It is direct contact with the contained product.
Principle of canning	Destruction of spoilage organisms within the sealed container by means of heat.
Prions	Smallest infective forms of life.
Probes	Different instruments used for instrumentation and their control (sensitive).
Probiotics	Probiotics are live microbial feed supplements that are beneficial to health (e.g. curd, sauerkraut etc.) Probiotics are living micro-organisms that following ingestion from part of the colonic flora at least temporarily and are used with a view to improving the health and well being of the host, e.g., Lactobacillus strain GG.
Processing Aid	Substances aused as a manufacturing aid to ehance the appeal or utility of a food or component. Includes clarifies, clouding agents, catalysts, flocculants, filter aids, crystallization inhibitors, etc.
Product development	It is an important activity in fod industry includes systematic approach to develop a new product through set of activities required to bring a new concept to a state of market readiness.
Productivity	it is concerned with reletion that exist between the aggregate contribution of production (land, labour, capital, and organization) result achieved by their continuation in an enterprise (productivity-output/ input, P=O/I).

Term	*Terminology*
Proffing	When the dough is allowed to site in a warm condition. The dough expands and risen because of yeast fermentation and production of CO2. The gas is trapped within gluten strnds.
Proof gallon	Proof gallon is a U.S. gallon of proof spirits or the alcoholic equivalent thereof, i.e., a U.S. gallon 3785 cm3 (231 cubic in.) containing 50% of ethyl alcohol by volume. Thus a gallon of liquor at 1200 proof is 1.2 proof gallons; a gallon at 860 proofs is 0.86 proof gallons. A British and Canadian proof gallon is an empirical gallon of 4546 cm3 (277.4 cubic in.) at 1000 proof (57.1% of ethyl alcohol by volume). An empirical gallon is equivalent to 1.2 U.S. gallons. To convert British proof gallons to U.S. proof gallons, multiply by 1.37. Since excise taxes are paid on the basis of proof gallons, this term is synonymous with tax gallons.
Proof	The alcoholic concentration of beverage spirits is expressed in terms of proof in Canada, the United Kingdom, and the United States. U.S. regulations defines this standards as follows: proof spirit shall be held to be that alcoholic liquor which contains one-half its volume of alcohol of a specific gravity of 0.7939 at 15.6°C i.e. the figure for proof is always twice the percent alcohol content by volume. For example, 1000 proof means 50% alcohol by volume. In the United Kingdom as well as Canada, proof spirit is such that at 10.6°C alcohol weighs exactly twelve-thirteenths of the weight of an equal bulk of distilled water. A proof of 87.70 indicates an alcohol concentration of 50%. A conversion factor of 1.142 can be used to change British proof to U.S. proof.
Proofing	Process of exposing dough to 31-32°C with 88% relative humidity in closed proofing chamber to increase loaf volume of dough by stimulating process of fermentation.
Pro-oxidants	Chemical substances and basic natural conditions or environmental states, which actively promote the oxidation of fats.

Term	Terminology
Propagation	Multiplication of plants by sexual or asexual means is called as propagation.
Propellant	Ga used to supply force to expel a product or used to reduce the amount of oxygen in contact with the food in packaging.
Protein association	Protein association reactions generally refer to changes occurring at the subunits or molecular level.
Protein calorie malnutrition (PCM)	Clinical and biochemical disorders caused by various degrees of deficiency and additional physical insults and stress. Inadequate dietary intake of good quality protein and calories.
Protein efficiency ratio (PER)	A measure of protein quality determined by the ability of a protein to support the growth of a young animal.
Protein isolate	Percent protein content is more than 90%. It is protein rich product obtained from oilseeds meal containing 85 to 90% protein.
Protein	A three-dimensional biological polymer constructed from a set of 20 different monomers called amino acids.
Protein-energy malnutrition (PEM)	This results when a person regularly consumes insufficient amounts of kcalories and protein. The deficiency eventually results in body wasting and an increased susceptibility to infections.
Prothrombin	A blood protein needed for blood clotting that requires vitamin K for its synthesis.
Proton pump	An active transport mechanism in cell membranes that consumes ATP to force hydrogen ions out of a cell and in the process generates a membrane potential.
Psychometric chart	A psychometric chart is a graph of the properties of air-water vapour mixtures as a function of temperature. Lines of constant wet bulb in psychometric chart are a plot of equation. If the 'wet bulb' temperature T_w and the dry bulb temperature T_a are known it is possible to establish the humidity H_a.

Term	Terminology
PUFA	Poly unsaturated fatty acids, more in vegetable oil.
Pungency	Certain compounds found in several spices and vegetables cause characteristic hot, sharp, and stinging sensations that are known collectively as pungency.
Quality	Composite of characteristics or degree of excellence.
Quota	A quota amount trade is a type of trade barrier that nations place on the physical amount of importas or exprots any of specific kind of goods.
Radiation dose	The quantity of radiation energy absorbed by a food as it passes through in processing.
Radiation	A hot body gives off heat in the form of radiant energy, which is emitted in all directions. When this energy strikes another body, part is reflected and part may be transmitted unchanged through the body, depending on its degree of opacity. The remainder is absorbed and quantitatively transformed into heat except where photochemical or special reactions are induced.
Radioactivity	It is property of an atom whose nucleus or center is physically unstable and spontaneously releases radiation energy.
Radurisation	Irradiation treatment is given for prologation of storage life by reducing vegetative bacteria.
Rancidity	Development of off flavor in food items due to break down of unsaturated fatty acids to saturated one.
	The development of off-flavour in stored oil/fat due to lipolysis or autoxidation.
Rate of flow	It is the amount of fluid that flows past a given point at any given instant.
Reactor	Reactor is a well designed vessel used for different bioconversion reaction.

Term	Terminology
Recommended dietary intake (RDI)	Recommendations from the original tenth edition RDA Committee that were published in 1987 in the American journal of Clinical Nutrition after the National Academy of Sciences refused to publish the original tenth edition of the RDA.
Recommended Nutrient Intake (RNI)	The Canadian version of RDA.
Reduced fat	25% or less fat than a full-fat product per serving.
Reduced or fewer calories	Any substance capable of absorbing heat from another required substance.
Reducing agent	In one sense a compound capable of donating electrons (also hydrogen ions) to another compound.
Reduction	To gain an electron or hydrogen atom.
Reference Daily Intake (RDI)	Standards of expressing nutrient content on nutrition labels. RDI figures are based on average 1989 RDA values set for a nutrient that span a particular age range, such as children over 4 years through adults. RDI soon should replace U.S. RDA.
Refractive index	A measure of the bending or refraction of a beam of light on entering a dense medium. It is the ratio of the sine of the angle of incidence of the ray of light to the sine of the angle of refraction.
Refrigeration	If the primary purpose is to discharge heat to a certain high temperature region, the system is called a heat pump. If the purpose is to absorb heat from a certain low temperature region, the system is called a refrigerator.
Regulated market	In these markets business is done accordance with rules and regulations.
Reichert-Meissel value	The R-M value of a fat is the number of milliliters of 0.1 N potassium hydroxide required to neutralize the stem-volatile fatty acids in 5 g of fat.
REM sleep	A periodic state during sleep identified by rapid eye movements and correlated with vivid dreaming.
R-enamel cans	Lacquered cans used for acid are known as R-enamel cans.

Term	Terminology
Rendering	Steam heating to oil seeds in vacuum container.
Renin	An enzyme formed in the kidney in response to low blood pressure; it acts on a blood protein to produce angiotensin I.
Repetitive DNA	DAN sequences those are present in a genome in multiple copies, sometimes a million tims or more.
Respiration rate	It is post harvest phenomenon where grains take in O_2 and give out CO_2.
Respiration	Fundamental process of conversion of potential energy to kinetic energy.
	Oxidative breakdown of the complex materials of cell into simples molecules with the concurrent production of energy required by the cell for the completion of chemical reactions.
Respiratory quotient	It is an arithmetic ratio of mole of CO_2 evolved to the moles of O_2 absorbed.
Restoration	Addition to restore the original nutrient content.
Restriction endonuclease	For molecular both the source DNA that contains the target sequence and the cloning vector must be consistency cut in to discrete and reproducible fragment.
Retention time	Time required coming out of eluent.
Retina	The innermost layer of the vertebrate eye containing photoreceptor cells (rods and cones) and neurons; transmits images formed by the lens to the brain via the optic nerve.
Retinal	The light-absorbing pigment in rods and cones of the vertebrate eye.
Retrogradation	The insolubilizating effect, initiated when the long and somewhat unwieldy molecules begin to crystallize is retrogradation when it occurs in starch.
Reversible process	System passes through a continuous series of a equilibrium state during a process.
Reversion	The development of a characteristic beany, buttery, fishy, grassy, haylike, or painty lavour in oil before the onset of autoxidation.

Term	Terminology
Reynolds number	The ratio inertial force to viscous force.
Rheology	It is viso-elastic property that studies deformation of dough including elasticity and flow.
	Rheology is concerned with stress-strain relationships of materials that show lavour r intermediate between those of solids and liquids.
Rhodopsin	A protein involved in vision; it is made in the eye and incorporates a protein called opsin and a form of vitamin A; especially important in night vision.
Rickets	A disease characterized by softening of the bones because of poor calcium deposition. This deficiency disease arises from lack of vitamin D activity in the body.
	The non-abnormalities due to deficiency of Ca, P and vitamin D.
Rickettsias	Intracellular parasites, gram negatives and only multiply only within host cells by binary fission.
Rigid packages	Packages made from strong material and having definite shape.
Rigour mortis	Stiffening of muscles due to loss of ATP and fall in pH during muscle postmortem.
Ripening	It is defined as the sequences of changes in colour, flavour, and texture which lead to state at which the fruit is acceptable to eat.
Roasting of groundnut	Dry heat treatment to groundnut for desirable flavor, light roasting.
Ropyness of bread	It is bread fault where sticky, gummy material is developed in center of loaf 1 to 3 days after baking due to contamination of microorganism B. mesentericus.
Royal jelly	It is used by bees to include development of larvae into the queen phenotype and is currently sold as a general health tonic
Saccharin	An alternate sweetener that yields no energy to the body; it is 500 times sweeter than sucrose.
Sake	Sake is undistilled alcoholic beverage produced from fermentation of rice having alcohol per cent about 15-17%.

Term	Terminology
Salt	Generally refers to a mixture of sodium and chloride in a 40:60 ratio.
Salty taste	Only salts are salty; however, not all salts are salty. Some are sweet, bitter, or tasteless. Monovalent cations, especially sodium, can pass directly through ion channels in the tongue, leading to an action potential leading to the salty percept. Sodium chloride in foods may be analyzed using a specific ion electrode. To measure other salts, ion chromatography or atomic absorption emission spectroscopy are generally used.
Sapid	Having the power of affecting the taste receptor.
Saprophyte	Saprophytes are those which live on dead and decay matter.
Sauce	Sauces are mainly two kinds; thin and thick sauces. The thin sauce contains vinegar, extract of spices and herbs. Thick sauce is more viscous and does not flow easily. It has at least 3% acetic acid for improving the keeping quality. Tomato pulp without skin and seeds are used as basic raw material for tomato sauce. Apples, peaches, plums, apricots, mangoes, cauliflower, carrots are also used for preparation of sauce. Onion, garlic, spices, and herbs are used as flavouring agents while vinegar; salt and sugar are added for taste. The mixtures of above all material cooked to get consistency of jam, i.e., upto 68 to 69°Brix.
Savory	Appetizing; having an agreeable flavour.
Scalding	It is process of treating birds with hot water and agitating for short period of time. It facilitates removal of feathers by expanding or relaxing muscles that surrounds feathers.
Scalding	Process of giving hot water treatment to facilitate removal of feathers.
Scale up	Scale up is the study of the problem associated with transferring data obtained in laboratory and pilot equipment to industrial production.

Term	Terminology
Schaal oven test	The sample is stored at about 65°C and periodically tested until oxidative rancidity is detected.
Scientist P. Fizer	In the year 1923 obtained 1st successful plant of citric acid fermentation with the help of sugar and M.O. Aspergillus niger.
Score	Which are used to represent their achievement are average scores of the two or more examinations.
SCP	Single cell protein, i.e., protein derived from microbial cell for use as food or a feed supplement.
Screening	The programme which consist of highly selective procedure used for detection and isolation of industrial important microorganisms.
Scurvy	The deficiency disease that results after a few weeks of consuming a diet free of vitamin C; pinpoint hemorrhages on the skin are an early sign.
Scutellum	It is cementing layer which separates endosperm from germ.
Secondary food processing	It is a step after primary processing where product is ready for consumption or another use. *e.g.* wheat milling.
Secondary response	Antibody made on second (subsequent) exposure to antigen mostly of class Ig G.
Sedimentation value	The amount of sediments deposited in lactic acid in 30 second in a measuring cylinder.
Sedimentation	Separation of a insoluble solid from a liquid in which it is suspended by settling under the influence of gravity or centrifugation
Seed material	The pure mycellial vegatative growth on sterlized grain material that can also be called as seed material.
Self-fertile	It refers to ability of plant to mature viable seeds without aid of pollens from some other plant flower.
Semi perishable foods	Food products which contain water in the range of 10-30% are called semi perishable food.
Semolina	Coarsely ground white endosperm chemically same to white flour.

Term	Terminology
Senescence	It is define as period during which anabolic (synthetic) biochemical processes give way to catabolic (degradative) processes leading to aging and finally the dead of tissue.
Sensory analysis	The science of measuring and evaluating the properties of food products by one or more human senses.
Sensory evaluation	Sensory evaluation is a scientific discipline that analyses and measures human responses to the composition of food and drink, *e.g.* colour, appearance, touch, odour, texture, temperature, taste and overall acceptability.
Septicemia	Infection of the bloodstream by micro-organisms.
Sequesterant	Substance, which combines with polyvalent metal ions to form a soluble metal complex to improve the quality and stability of product.
	Chemical substances used to chelate or complex free metals in food system.
	These are chelating agents or sequestering compounds. They react with trace elements such as iron and copper present in foods and remove them from solution. The trace elements are active catalysts of oxidation and discolouration of food products. Sequestrants such as ethylene diamine tetra acetic acid (EDTA), poly phosphates and citric react with trace elements and inactivate them.
Serbet (Juice)	It is clear sugar syrup, which has been artificially flavoured.
Settling down period	Period at which solid surface conditions comes into equilibrium with drying air.
Settling	It is a method of oil refining. It involves storing heated fats quiescently in tanks with conical bottoms. Water and materials associated with water settle into the cone from where they are drawn off.
Sherbats	This is cooling drink of sweetened diluted fruit juice.

Term	Terminology
Sherry	It is term given to number of related type of dessert wine originally developed in the area around Serez in the South of Spain.
Shortenings	These are compounded from mixtures of fats prepared by hydrogenation and are called lard compounds or lard substitutes.
Sifter	Machine used for sieving impurities smaller or larger than paddy grains.
Simultaneous reaction	Nutrients converted to products in a variable proportion without accumulation of intermediate
Single point recorder	The instruction that traces the chanages of one measured variable are called single point recorder.
Single whiskey	Single whiskey is the whiskey, either grain or malt, produced by one particular distillery. Blended Scotch whiskey is not a single whiskey.
Size reduction	The break down of solid materials through the application of mechanical forces.
Smoke point	The temperature at which the column of solidified oil or fats starts rising in the capillary tube.
Smoking	Decreases the available moisture on surface of the meat, preventing microbial growth and spoilage. It enhances colour and falvour of meat.
Soaking of legume	Absorption of water/medium through hull or seed coat.
Sodium-potassium pump	A special transport protein in the plasma membrane of animal cells that transport sodium out of and potassium into the cell against their concentration gradients.
Softening point	The temperature at which the column of solidified oil or fat strats rising in the capillary tube.
Soil	It can be defined as superficial earth crust which functions as store lower time providing necessary physical support to plant.
Solubility	The tendency of molecule to dissolve in a liquid.
Solute	A substance that is dissolved in a solution.
Solvent or Vechicle	Substance used to extract or dissolve another substance.

Term	Terminology
Solvent	The dissolving agent of a solution. Water is the most versatile solvent known.
Sour mash	Sour mash is made with a lactic culture and not less than 20% stillage added back to the fermentor and fermented for at least 72 h.
Sour taste	Sourness indicates acidity, through not all acids are sour. The detection of acids facilitates maintaining the body fluid compositional balance. The pH is characteristic of the carbon dioxide levels in blood and cerebrospinal fluid.
Specific gravity	The density of a substance is defined as mass per unit volume. The density of a substance is characteristic property and has a definite value at a given temperature and pressure. The density of one substance in relation to the density of another material (e.g., water) is known as specific gravity.
Spencer's theory	According to this theory pectin is negatively charged. In jelly formation, sugar acts as a precipitating agent and the presence of acid helps it.
Spirits	Spirits are distilled spirits including all singular whiskey, gin, brandy, rum, cordials, and others made by a distillation process for non-industrial use.
Spore	Resistant, resting stage of bacteria.
Spray drying	Spray drying is usually applied to fluids high in moisture content. The fluid is divided into droplets by its passage through a spray nozzle at high pressure or its passage through a centrifugal disk at high speed.
Springiness	The elastic recover that occures when the compressive force is remove.
Sprouting of legumes	Soaking of legumes in water and then kept for vegetative growth.
Squash	It is essentially strained juice containing moderate quantities of fruit pulp to which cane- sugar is added for sweetness.

Term	Terminology
Stabilizer or Thickener	Substance used to produce viscous solutions or dispersions, to impart body, improve consistency, or stablize emulsions. Includes suspending and bodying agents, setting agents, bulking agents.
Stabilizers	These are the compounds function to improve and stablize the texture of foods, stabilize emulsion and foams, inhibit crystallization and encapsulate flavor.
Starch	Starch referred as plant glycogen.
Starter culture	Culture of desirable microbes used to initiate fermentation.
Stature	Height of the body in a standing position.
Stepwise reaction in kinetic pattern	Nutrients completely converted to intermediate before conversion to product.
Sterilization	Sterilization means the destruction of all viable microorganisms by heat treatment.
Steroids	A group of hormones and relate compounds that are derivatives of cholesterol.
Stevens power law, S=In	The increase in perceived intensity, S, is equal to the concentration, I, to the nth power.
Stillage	Stillage is dealcoholized fermented mash.
Strecker degradation	It involves the interaction of -dicarbonyl compounds and -amino acids. Volatile products, such as aldehydes, pyrazines and sugar fragmentation products from the strecker reaction may contribute to aroma and flavour.
Stress	A pattern of disruptive psychological and physiological functioning that occurs when an environmental event is appraised as a threat to important goals and one's ability to cope.
Stunning	Of making animal unconscious before slaughtering.
Subjective evaluation	Human sense organs for taste, smell, sight, touch and hearing are the ways for subjective evaluation of food.
Sugar alcohol	Sugar contain with alcohol derviative e.g. Sorbitol.

Term	Terminology
Superheated steam	When dry steam is further heated at constant pressure thus raising its temperature.
Superoxide dismutase	An enzyme that can neutralize a superoxide free radical.
Surface Finishing Agents	Substance used to increase palatability, preserve gloss and inhibit discoloration of foods. Includes glazes, polishes, waxes and protective coatings.
Surface-Active Agent	Substance used to modify surface properties of liquid food components for a variety of effects, other than emulsifiers. Includes solubilizing agents, dispersants, detergents, wetting agents, rehydrating ehancers, foaming agents, defoaming agents, etc.
Sweating	Moisture condensation on surface of endosperm during wheat conditioning.
Sweet taste	Two types of receptors systems correspond with sweet taste; one responds to certain carbohydrates and the other to high potency sweeteners. The structural requirements for a compound to active sweet receptors have not been fully defined. High pressure liquid chromatography is a key tool used in analyzing sweet components.
Sweeteners	These are substances other than sugar added to keep the food energy low or they have beneficial effects for diabetes mellitus and tooth decay and are added for flavouring of foods.
Syneresis	The phenomenon of spontaneous exudation of fluid from gel.
Synergist	Substance used to act or react with another food ingredient to produce a total effect different from or greater than the sum of the effects produced by the individual ingredients.
Syrup	It is clear sugar syrup, which has been artificially flavoured. It contains 25-30% juice, 59% sugar and 1% citric acid. Syrup used as a sherbet by diluting 1 cup of syrup with 3 cups of water.
Tails	Tails is a residual alcoholic distillate.
Tallow	It is a lipid obtained from beef by the process of rendering.

Term	Terminology
Tannins	The phenolics with MW 500 to 3000 that form complexes with proteins.
Taste	Chemoreceptive events in the mouth lead to taste perception. Taste is typically described by five modalities coupled with chemesthesis: salt, sour, sweet, bitter and umami.
TBA (thibarbituric acid) test	Thiobarbituric acid is a more sensitive reagent than phloroglucinol and gives a yellow coloration with various saturated and unsaturated aldehydes and a red colour with a more restricted number.
TD (true digestibility)	N intake - (fecal N - metabolic N) x 100 ÷ N intake.
TDP	It is the temperature necessary to kill the organisms in 10 minutes.
Teloentric	Centromere is present at the proximal end on a rod like chromosome.
Temperature	It is defined as the degree of hotness or coldness measured on a definite scale.
Tempering of chocolate	It is the most important process in chocolate manufacturing which decides the quality. It consists of cooling the chocolate with continuous mixing to produce coca butter seed crystals and uniform distribution it is a technique of uniform distribution.
Tempering	Process of equalizing the moisture of grain by temporary holding the paddy/any grains between drying passes.
Ten State Nutrition Survey (TSNS)	A survey, which was funded by the Department of Health, Education and Welfare (United States) to study the problem of malnutrition in the United States.
Tetany	A syndrome marked by sharp contraction of muscles with failure to relax afterward; usually caused by abnormal calcium metabolism.
Texture	Closely related to the rheology of food materials the texture of food. It is a structural quality of a material.

Term	Terminology
Texturizer	Substance, which affects the appearance or feel of the food.
	The chemical substances used to improve texture of food products.
	These are the substances used to improve the texture of processed fruits and vegetables that would remain healthy otherwise become soft.
Thermal death time	The time of rexposure to heat during which original number of viable microbes is reduced by one tenth.
Thermal reactions	Thermal reactions occurring with carbon-carbon bond cleavage yield as primary products volatile acids, aldehydes, ketones, diketones, furans, alcohols, aromatics, carbon monoxide and carbon dioxide.
Thermodynamics	Thermodynamics is the science of energy transfer in relation to the physical properties of substances.
Thermoplastic	The plastics which can be melt and solidify agains and again.
Thickeners	The chemical substances used to improve thickness, viscosity and water holding capacity of liquid or semi-liquid foods.
Thiobarbituric acid (TBA) test	This is one of the most widely used tests for evaluating the extent of lipid oxidation. Oxidation products of unsaturated systems produce a colour reaction with TBA. It is believed that the chromagen results from condensation of two molecules of TBA with one molecule of malonaldehyde.
Thiocyanogen value	This value is the result of determining the percentage of oleic, linoleic and linolenic acids, provided that other unsaturated acids are not present and the iodine number and the amounts of saturated acids are known.
Titer test	When molten fatty acids are cooled and begin to solidify, the latent heat of fusion is liberated and a sudden rise in temperature can be observed.
Tocopherols	The chemical name for some forms of vitamin E.

Term	Terminology
Tomato juice	It is a juice, which contains T.S.S. not less than 5.6% at 20°C (4-6% salt is added to counteract the astringent taste of the juice and 1% sugar is added to improve the taste).
Tomato ketchup	It is a product made by concentrating tomato juice or pulp (without seed and pieces of skin). Spices, salt, vinegar, onion, garlic are added to the extract that the ketchup contains not less than 12% tomato solids and 28% total solids.
Tomato paste	It is a concentrated tomato juice (without skin and seeds) containing not less than 25% tomato solids. It is further concentrated to 33% tomato solids.
Tomato puree	It is a concentrated tomato pulp without skin or seeds containing not less than 8.37% tomato solids. It is further concentrated to not less than 12% tomato solids. It is called heavy purees.
Top yeast	The clumping of the cells and the rapid evolution of CO_2 sweeps the cells to the surface hence termed as top yeasts.
Total and volatile carbonyl compounds	Methods for the determination of total carbonyl compounds are usually based on measurement of the hydrazones arising from reaction of aldehydes and ketones (oxidation products) with 2, 4-dinitrophenylhydrazine.
Total fiber	Total fiber as the sum of dietary fiber and functional fiber.
Totipotency	Capacity of cell to regenerate phenotype of the complete and differentiated organism from which it is derived.
Toxic syndrome	The toxicity syndrome, which result from ingesting food stuffs which has been inadvertently contaminated with certain modl.
Toxin	Any harmful chemical substance released by microbes.
Tracer	Substance added as a food constituent (as required by regulation) so that levels of this constituent can be detected after subsequent processing and/or combination with other food materials.

Term	Terminology
Transcription	The process of synthesis of mRNA over the DNA.
Transduction	Bacteria phases function as intermediate in transfer of bacterial genetic information from one bacteria to another.
Transgene	Transgene means the foregin DNA that is introducting into the genome.
Transient transfection	The introduced gene is retained and expressed in all the cells derived from the transfected cells.
Translation	The process of synthesis of polypetide from mRNA
Transpiration	It is important physiological activity of fruit and vegetables in which water loss from lenticels or pores to surrounding atmosphere takes place relates to physiological loss in weight.
Tricep Skinfold Thickness	Measurement, with a calibrated caliper, of thickness of a fold of skin at the upper arm. It is a measurement of subcutaneous fat. Since subcutaneous fat make up about 50% of the adipose tissue stores, tricep skinfold thickness can be an important measurement to estimate body fat reserves of an individual.
Trios phosphate	It is an intermediate product of glycolysis.
Turbidity point	The temperature at which a solution of oil or fat starts showing turbidity when it is cooled slowly.
Turbulent flow	The flow in which the fluid instead of flowing in orderly manner moves erratically in the form of cross and eddies is called turbulent flow.
Ultimate pH	The lowest pH of muscles after slaughter of animal.
Ultra filtration membranes	Membrane is preventing the passage of larger solute molecules in solution by means of filtration through micro pores in a membrane structure.
Ultra filtration	The process of separating components of a solution largely on the basis of molecular size, utilizing a membrane as a molecular sieve.
Umami taste	Umami is the taste of a few amino acids (e.g., glutamate, aspartate, and related compounds) and was classically not included as a taste modality. Sometimes described as savory, brothy, or meaty,

Term	Terminology
	it is the dominant taste of such foods as chicken broth, meat, and ageing cheese. Umami perception results from activation of two receptor systems, with one that overlaps with the receptor systems for artificial sweeteners. There are many hplc systems that integrate sample preparation along with data analysis specific to amino acids analysis.
Unbalanced transportation problem	The transportation problem in which the requirement and capacity are not balanced.
Under nutrition	Inadequate intake of one or more nutrients including calories. Chronic under nutrition (chronic malnutrition) refers to a long-term inadequate food intake and is reflected by low height-for-age levels. Acute under nutrition refers to a short-term severe inadequate food intake and is reflected by low weight for height levels.
	State of insufficient supply of essential nutrients.
Under weight	Weight below the average for a given sex, age and height in relation to the NCHS reference tables.
Unit operation	It is industrially design operation without change in chemical nature.
	Aerobic fermentation involves unit operation the mixing of three heterogeneous phase microorganisms, medium and air mass transfer of oxygen from the air to the organisms and heat transfer from the fermentation medium.
Unit process	It is a combined process in which both physical and chemical changes are involved.
	Fermentation processes can be classified by the reaction mechanisms involved in converting the raw materials into products; these include reductions, simple and complex oxidations, substrate conversions, transformations, hydrolysis, polymerization, complex biosynthesis and formation of cells.
Unsaponifiable matter	This term includes all those constituents of fats, which are not saponified by alcoholic caustic potash. They include sterols, carotenoids, long-chainmatter alcohols and hydrocarbons.

Term	Terminology
Unsaturated fatty acids	Hydrocarbon chain of fatty acid contains double bonds between carbon atoms.
Uric acid	An insoluble precipitate of nitrogenous waste excreted by land snails, insects, birds and some reptiles.
UV Spectro-photometry	Measurement of absorbance at 2 nm (conjugated dines) and 268 nm (conjugated trienes) is sometimes used to monitor oxidation.
Variability	In a frequency distribution, the separation, dispersion, or spread of the scores on the x-axis.
Variable cost	The cost which varies with the amount produced, raw material, labour.
Vector	A DNA molecule known to replicate autonomously in a host cell, to which a segment of DNA may be splied to allow its replication.
Vegetable oil	Vegetable oils mostly contain large amounts of oleic and linoleic acids and less than 20% saturated fatty acids. The most important group is cottonseed, corn, peanut, sunflower, safflower, olive, palm, and sesame oils.
Verification	Routinely check the system for accuracy to verify that it is functioning properly and consistently.
Very low calorie diet (VLCD)	Known also as protein sparing, modified fast (PSMF), this diet allows a person 400 to 700k calories per day, often in liquid form. Of this, 30 to 120 grams are carbohydrate; the rest is high biological value protein.
Very low density lipoprotein	The lipoprotein that initially leaves the liver. It carries both the cholesterol and lipid newly synthesized by the liver.
Vinegar	It is defined as the condiment made from sugary or starchy materials by an alcoholic fermentation followed by an acetous one.
Virulence	It is the degree or capacity or intensity of pathogencity of an organism, i.e., to cause the disease in host.

Term	Terminology
Virus	The viruses are submicroscopic entities capable of being introduced into specific living cells and reproducing inside such cells only.
Viscosity	The resistance to flow of a fluid. In the cgs (centimeter-gram-second) system, the absolute unit viscosity is the poise.
Vitamin	An organic molecule required in the diet in very small amounts; vitamins serve primarily as coenzymes or parts of coenzymes.
	Organic compounds occurring in small quantities in the different natural foods and necessary for the growth and maintenance of good health in human being.
Viticulture	It can be defined as the science and art of growing grapes.
VOC process	It represents the voice of customers and is proactive way to capture the changing requirements of the customers with time.
Volatibility	The tendency of a molecule to pass into the vapour state.
Volatile oil	That portion of a botanical that co-distills with water during steam distillation and is generally flavourful.
Wash	Wash is the liquid obtained by fermenting wort with yeast. It contains the beverage spirit and congeners developed during fermentation.
Washing or Surface Removal Agent	Substance used wahses or assist in the removal of unwanted surface layers from plant or animal tissues.
Water activity (aw)	F/Fo = The fugacity of the solvent/The fugacity of the pure solvent.
	It is a ratio between the partial pressure of water in a sample (P) to the vapor pressure of pure water at the same temperature (Po): aw = P/Po.
	Percent equilibrium relative humidity surrounding the product.

Term	Terminology
Water holding capacity	It is a term frequently employed to describe the ability of a matrix of molecules usually macromolecules to entrap large amounts of water in a manner such that exudation is prevented.
Water potential	The physical property predicting the direction, in which water will flow, governed by solute concentration and applied pressure.
Water structure	Li^+, Na^+, H_3O^+, Ca^{2+}, Ba^{2+}, Mg^{2+}, Al^{3+}, F^- and OH^-.
Water structure breakersions	K^+, Rb^+, Cs^+, NH_4^+, Cl^-, Br^-, I^-, NO_3^-, BrO_3^-, IO_3^-, and ClO_4^-
Wavelength	In measuring light and sound waves, the distance from the crest of one wave to the crest of the next.
Waxes	Esters of long-chain monohydroxy alcohols and higher-chain fatty acids.
Waxy rice	Rice with low amylose content
Weber's law	Principle stating that the amount by which a stimulus must be increased or decreased to be perceived as different is always a constant proportion of the initial stimulus intensity.
Wet milling of legume	Related to pre-milling treatment, soaking in water.
Wheat shorts	Mixture of bran and germ produced from milled wheat.
Whey	It is by product of cheese and casein and contains approx. 20% of the original milk proteins. These proteins include lactalbumin, lacto globulin, lactoferrin, lactoperoxidase, immunoglobulins, glycomacropeptide and a variety of growth factors.
Whitening of rice	The process of removal of bron from brown rice called as whitening of rice.
Whole foods	Whole foods are unprocessed and unrefined, or processed and minimally refined, before being consumed (*e.g.* wheat porridge).
Wine Gallon	Wine gallon is the measure of actual volume; a U.S. gallon (3.785 L) contains 3785 cm3 (231.0 cubic in.) British (Imperial) gallon contains 4546cm3 (277.4 cubic in.)

Term	Terminology
Winterizing	A process where oil is held at 5°C until crystallization is well advanced, and then filtered in chilled room to remove solids. The winterized oil is used as salad oil.
Work	A force is a means of transmitting an effect from one body to another.
Wort	It is the clear liquid obtained through the filtration of insoluble materials during mashing process of brewing.
Writing	One of the ways through which we express our thoughts is writing.
Xantham	It is polysaccharides produced by Xanthomonas compestris.
Xerophthalmia	A cause of blindness that results from infection of the eye secondary to vitamin A deficiency. The specific cause is a lack of mucus production by the eye, which then leaves it more vulnerable to surface dirt and bacterial infections.
Xylitol	An alcohol derivative of the five-carbon monosaccharide, xylose.
Yeast	A unicellular fungus that lives in liquid or moist habitats, primarily reproducing asexually by simple cell division or by budding of a parent cell.
	Single cell micro organism which multiplies by process of budding. They are single cell protein and rich source of enzymes like lipase, amylase, protease and zymase.
	Fungi having a predominantly unicellular organisation and are collectively and popularly called yeasts.
Yogurt	It contains live lactic acid bacteria (probiotic).
Yolk index	Height of yolk/breadth of yolk.
	It is used to evaluate egg quality height of the yolk in relation to width of yolk. Yolk index is high in fresh egg.

Term	Terminology
Yo-Yo dieting	The practice of losing weight and then regaining it, only to lose it and regain it again.
Z value	Number of degree required for a specific thermal death time curve to pass through one log cycle.
Zeroth law of thermo-dynamics	If two systems are both in thermal equilibrium with third system, they are in thermal equilibrium with each other.
Z-Score	A score's distance from the mean of the group expressed in units of the standard deviation; used to identify unusual values.
Z-Value	It is interval in temperature at 0°F required for a line to pass through one logarithmic cycle on the semi logarithmic paper.
Zwitter ion	Amino acid having net charge zero, i.e., equal no positive and negative charges.
Zymogen	An inactive form of an enzyme.

4

DISTINGUISH / COMPARISON

Distinguish between Vitreous and Mealy Wheat

Vitreous wheat	Mealy wheat
1. It includes Triticum dicoccum, Triticum monococcum, Triticum durum, Triticum aegiloploids	It includes Triticum turgidum, Triticum compactum, Triticum aestivum
2. Not favoured by heavy rainfall, light sandy soil, crowded planting	Favoured by heavy rainfall sandy soil, crowed planting
3. Favoured by nitrogenous fertilizers	Not favoured by nitrogenous fertilizers
4. Positively correlated with protein content	Positively correlated with high grain yielding capacity

Distinguish between Hard and Soft Wheat

Hard wheat	Soft wheat
1. Fragmentation of endosperm is long the cell boundaries	Endosperm fractures in a random way
2. Mechanically very strong	Mechanically very weak
3. Mr 15 K protein, i.e., Friabilin present in less amount or entirely absent	Mr 15 K protein, i.e., Friabilin present in more amount. It is non stick agent
4. Yield coarse flour with regular shaped particles	Yield fine flour with irregular shaped particles

Distinguish between Strong and Weak Wheat

Strong wheat	Weak wheat
1. High protein content	Low protein content
2. Ideal for bread making	Ideal for biscuit, cookies and cakes
3. Need not any blending	Need to be blend with strong flour to use for bread
4. Slejpner, Haven and Maris Huntsman are hard wheat not good for bread making	Minaret and flanders are soft wheat acceptable for bread making

Distinguish between Dry Milling and Wet Milling of Corn

Dry milling of corn	*Wet milling of corn*
1. Objective: Separation of kernel into its anatomical parts	Separation of corn into its chemical constituents
2. Fractions: endosperm grits bran, germ, hominy feed	Starch, protein, fiber, oil
3. Size of endosperm: more	Less
4. Shelf life: Non-degerming method has less life	More
5. Uses: breakfast cereals, fermented beverages, extruded snacks, 34% feed.	Corn oil, defatted germ meal
6. Grits of endosperm is the primary product	Starch is the primary product

Distinguish between Disc Shellers and Rubber roll Shellers

Disc shellers	*Rubber roll shellers*
1. Comparatively more breakage of grains	It causes less breakage of grains
2. Low efficiency	It has greater efficiency of mute removal
3. Horizontal emery covered disc	It has two rubber rolls
4. One is moving other stationary	Both rotate in opposite direction

Distinguish between Aseptic and Conventional packaging

Aseptic packaging	*Conventional packaging*
1. Food sterilized outside the packaging	Food sterilized inside the packaging
2. Energy saving is more	Energy saving is less
3. More hygienic	Less hygienic

5

SHORT EXPLANATIONS

WHAT IS FOOD SCIENCE?

Food Science has been defined as the application of the physical, biological and physcological sciences to the processing and marketing of foods. The main emphasis in food science is given to the technological and the nutritional aspects. It must be remembered that food is eaten primarily to satisfy the needs of the body for nutrients. The facts make it clear that food scientists should have a basic understanding of human nutrition and convert raw Agricultural and fishery products into nutritious as well as acceptable processed products.

The field of human nutrition, especially the aspects most relevant to food science and technology covers:

- The history of nutritional knowledge
- Individual nutrients and their physiological functions
- Nutrient content of foods
- The maintenance and even the improvement of nutritive values during processing, preservations and distribution.

WHAT ARE FOODS?

Foods are materials, raw, processed, or formulated, that are consumed orally by humans or animals for growth, health, satisfaction, pleasure, and satisfying social needs. Generally, there is no limitation on the amount of food that may be consumed. This does not mean that we can eat any food item as much as we want. Excessive amounts could be lethal, for example, salt, fat, and sugar. Chemically foods are mainly composed of water, lipids, fats and carbohydrate with small proportions of minerals and organic compounds. Minerals include salts and organic substances include vitamins, emulsifiers, acids, antioxidants, pigments, polyphenols and flavour-producing compounds. The different classes of foods are perishable, harvested, fresh, minimally processed, preserved, manufactured, formulated, primary, secondary derivatives, synthetic, functional and medical foods. The preservation methods are mainly based on the types of food that need to be preserved or formulated.

WHICH ARE THE FACTORS AFFECTING FOOD QUALITY, SAFETY AND CHOICE?

Following various factors is responsible for the food quality, safety and choice.

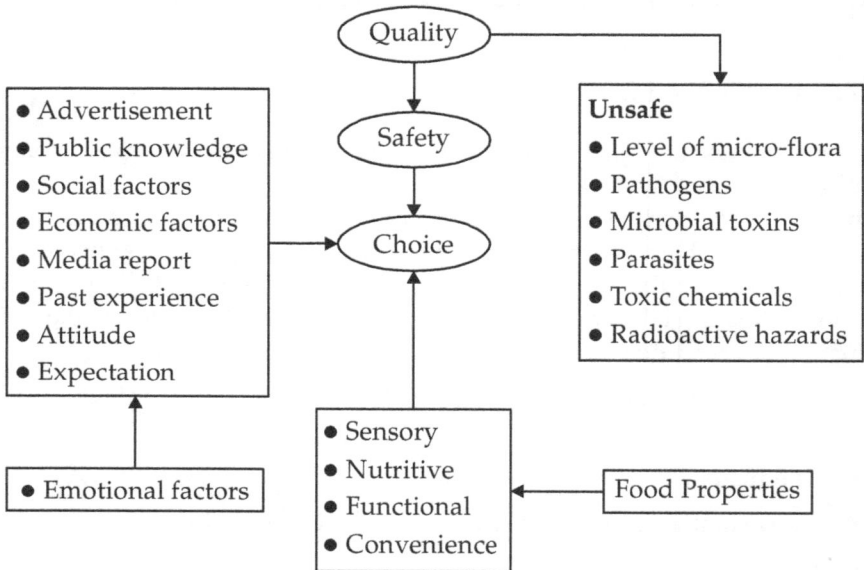

EXPLAIN FOOD EVALUATION IN SHORT?

Food evaluation is an important part of the process of developing new food products and of analyzing the market potential for these foods; likewise it is necessary in the study of processing and storage effects. Food evaluation provides information pertinent to the quality of food product involving chemistry, physics, technology of the food, its degree of excellence, which encompasses taste, aroma, texture, appearance and nutritional content as well as factors that determine the safety of food. Food evaluation system is divided in to two main groups 1. human sensory evaluation (subjectively evaluation) and 2. instrumental evaluation (objective evaluation).

The palatability of food may be judged on the basis of the kinds, quality and intensity of sensory impressions made. The most important sensory properties of food are flavour (comprising taste and odour), temperature (sensations of heat and cold), appearance and texture or mouth-feel (affecting the senses of touch).

Methods of sensory evaluation

    A.   Analytical or objective method.

    B.   Hedonic or subjective method.

A. Different tests under analytical method:

1. Single sample test (A note-A test)
2. Paired difference test
3. Triangle difference test
4. Duo-trio test
5. Ranking tests
6. Quality tests
7. Scoring tests
8. Descriptive tests
9. Flavour profile method
10. Dilution flavour profile method.

B. Different tests under hedonic method:

1. Preference tests
2. Paired preference test
3. Triangle preference test
4. Hedonic scale

Sensory evaluation is as scientific discipline used to evoke measure, analyse and interpret reactions to these characteristics of foods that are perceived by the senses. Texture measurement can be done using rheological techniques: Rheology deals with the deformation and flow of both liquids and solids (tenderness, firmness or softness).

What are the major functions of nutrients in human body?

Nutrient	Function(s)		
	Body building	Energy giving	Regulating
Carbohydrates	AF	MF	-
Proteins	MF	AF	AF
Lipids/Fats	AF	MF	-
Minerals	MF	MF	AF/MF
Vitamins	MF	AF	-
Water	MF	AF	MF

MF = Main function, AF = Additional function.

How much calorie and protein requirements of children?

Age group	Body weight (kg)	Daily energy need (Kcal)	Daily protein need (g)
Birth to 6 months	3 - 7	400-800	7 - 13
6 months - 1 year	7 - 9	800-900	14 - 17
1 - 3 years	10 - 15	1100-1400	19 - 22
4 - 6 years	16 - 22	1500-1900	29 - 30
7 - 9 years	24 - 30	1800-2200	36 - 42
10 -12 years	32 - 43	1900-2300	43 - 55
13 - 15 years	46 - 48	2060-2450	65 - 70
16 - 18 years	49 - 57	2060-2640	63 - 78

Ref: ICMR, New Delhi

Give the balance diet of 60 kg person (diet in g)?

Ingredient	Pregnant woman	Lactating woman	Normal working person	Growing children	
				Boys	Girls
Cereals	350	400	475	450	350
Pulses	45	55	65	60	50
Green vegetables	150	150	125	100	150
Other vegetables	125	125	175	175	150
Fruits	150	150	125	125	125
Milk/milk products	225	225	100	150	150
Oils/lipids	35	50	40	80	40
Eggs/meat/mutton	60	60	60	60	60
Sugar/Jaggery	40	50	40	40	30

What is the nutrient requirement per person/day?

Nutrient	Quantity/day
Carbohydrates	350 - 400 g
Proteins	60 g
Lipids/Fats	40 g
Calcium	400 mg
Iron	28 mg
Vitamin A	0.6 mg
Thiamin (B_1)	1.4 mg
Riboflavin (B_2)	1.6 mg
Pyridoxine (B_6)	2.0 mg

Nutrient	Quantity/day
Cobalamin (B$_{12}$)	0.05 mg
Niacin/Nicotinic acid	18 mg
Folic acid	0.1 mg
Vitamin C	40 mg

LIPIDS

Lipids are naturally occurring biomolecules comprising a large and diverse group of compounds that are saponifiable esters of long chain fatty acids. They are soluble in fat solvents such as chloroform, ether, hexane and benzene.

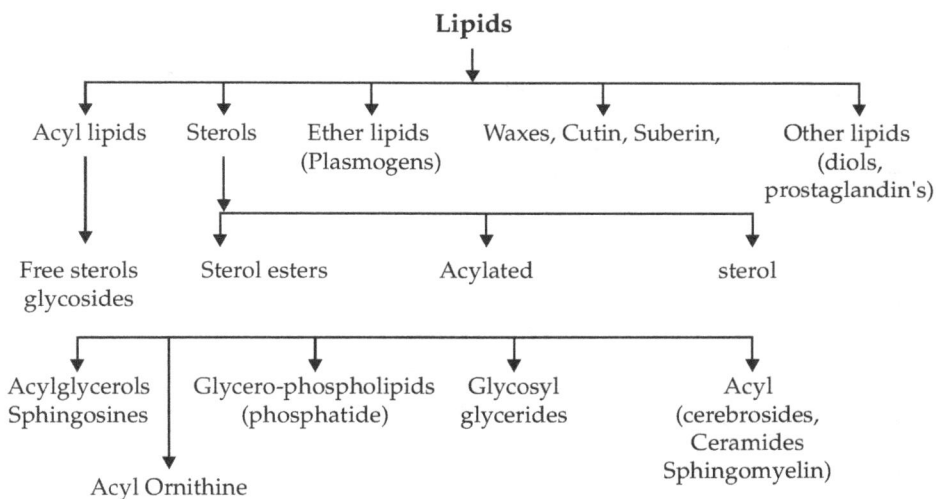

Lipids

Acyl lipids   Sterols   Ether lipids (Plasmogens)   Waxes, Cutin, Suberin,   Other lipids (diols, prostaglandin's)

Free sterols glycosides   Sterol esters   Acylated   sterol

Acylglycerols Sphingosines   Glycero-phospholipids (phosphatide)   Glycosyl glycerides   Acyl (cerebrosides, Ceramides Sphingomyelin)

Acyl Ornithine

Nomenclature of some common fatty acids

Common name	Systematic name	Abbreviation	Symbol
Butric	Butanoic	4:0	B
Caproic	Hexanoic	6:0	H
Caprylic	Octanoic	8:0	Oc
Capric	Decanoic	10:0	D
Lauric	Dodecanoic	12:0	La
Myristic	Tetradecanoic	14:0	M
Palmitic	Hexadecanoic	16:0	P
Stearic	Octadecanoic	18:0	Sta
Eicosanoic	Arachidic	20:0	Ad
Palmitoleic	9- Hexadecanoic	16:1	Po
Oleic	9- Octadecanoic	18:1 (9)	O
Linoleic	9,12- Octadecadienoic	18:2 (6)	L

Common name	Systematic name	Abbreviation	Symbol
Lenolenic	9,12,15- Octadecatrienoic	18:3 (3)	Ln
Arachidonic	5,8,11,14-Eicosatetraenoic	20:4 (6)	An
Erucic	13-Docosenoic	22:1 (9)	E
	Eicosapentaenoic acid	20:5 n3	EPA
	Docosahexaenoic acid	22:6 n3	DHA

Types of fatty acid oxidation:

β-Oxidation

α-Oxidation

ω-Oxidation

In each cycle of -oxidation 5ATP molecules are formed.

1. $FADH_2 \xrightarrow[\text{Formed during 1st oxidation steps}]{\text{Respiratory chain (oxidation)}} 2\ ATP$

2. $NADH \xrightarrow[\text{Formed 2nd oxidation steps}]{\text{Respiratory chain (oxidation)}} 3\ ATP$

AUTOXIDATION

Oxidations of lipids/fats occur in the presence of oxygen or enzyme or metals. Three steps involved in the oxidation process as follows.

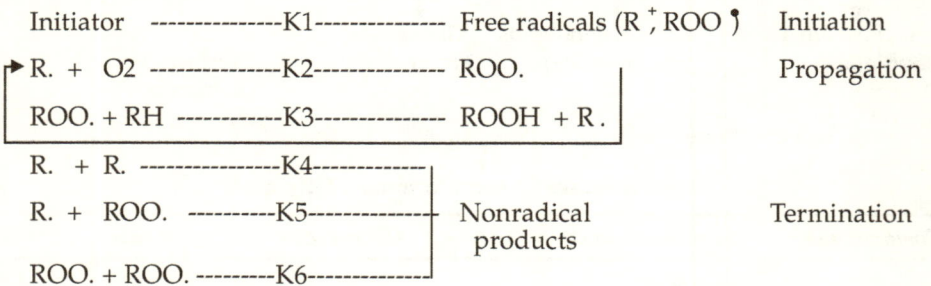

```
Initiator   --------------K1--------------- Free radicals (R ; ROO )    Initiation

R. + O2 ---------------K2--------------- ROO.                           Propagation

ROO. + RH ------------K3--------------- ROOH + R.

R. + R. ---------------K4------------

R. + ROO. ----------K5-------------  Nonradical                        Termination
                                      products
ROO. + ROO. ---------K6-------------
```

Electrons can exist in two different orientations, but with equal magnitude of spin, +1 and −1. The total angular momentum of the electrons in an atom is described by expression 2S + 1 where S is total spin. $2(1/2 + 1/2) + 1 = 3$ and $2(1/2 - 1/2) + 1 = 1$, i.e. ($^3O_2$) singlet stage. Singlet state oxygen is more electrophilic than triplet state oxygen. It can thus react rapidly (1500 times faster than $^3O_2$) with moieties of high electron density, such as C = C bonds. Several substances found in fat containing foods that can act as photosensitizers to produce $^1O_2$.

β-Carotene is the most effective $^1O_2$ quencher and tocopherols are somewhat effective. Synthetic quenchers such as BHA and BHT are also effective and permissible in foods.

The formation of hydroperoxides by singlet oxygen proceeds via mechanisms different than that for free radical autoxidation. Once the initial hydroperoxidies are formed, the free radical chain reaction prevails as the main mechanism. Each hydroperoxide produces a set of initial breakdown products that are typical of the specific hydropeoxide and depend on its position in the parent molecule. Hydroperoxides begin to decompose as soon as they are formed. In the first stage of autoxidation their rate of formation exceeds their rate of decomposition. The first step in hydroperoxide decomposition is scission at the oxygen-oxygen bond of the hydroperoxide group-giving rise to an alkoxy radical and hydroxy radical.

What types of rancidity occurring in fats, oils and fatty acids?

Type of rancidity	Main substances producing rancidity	Types of chemical reaction	Materials subject to the type of rancidity
Lipolytic	Low fatty acids, medium-chain fatty acids	Enzymic hydrolysis	Milk fat, plan-seed oils
Oxidative	Lower aldehydes and ketones	Autoxidation, enzymic oxidation	Polyunsaturated edible oils
Flavour reversion	Oxygen-substituted cleavage and rearrangement products	Oxidation, cleavage and rearrangement	Soybean oil
Ketonic	2-Alkanone (methyl ketones)	-Oxidation and enzymic decarboxylation	Milk fat, palm-seed oils

FACTORS INFLUENCING THE RATE OF LIPID OXIDATION IN FOODS

- Fatty acid composition
- Free fatty acids verses the corresponding acylglycerols
- Oxygen concentration
- Temperature
- Surface area
- Moisture
- Pro-oxidants
  - Acceleration of hydroperoxide decomposition
  - Direct reaction with the unoxidized substrate
  - Activation of molecular oxygen to give singlet oxygen and peroxy radicals.
  - Radiant energy
  - Antioxidants

ANTIOXIDANTS

Antioxidants are substances that can delay the onset or slow the rate of oxidation of autoxidizable materials.

- Tocopherols
- Gum guaiac
- Propylgallate (PG)
- Butylated hydroxyanisole (BHA)
- Butylated hydroxytoleuene (BHT)
- 2,4,5-Trihydroxy butyrophenone (THBP)
- 4-Hydroxymethyl-2,6-di-tert-butylphenol
- tert-Butylhydroquinone (TBHQ)

SYNERGISM

Synergism occurs when a mixture of antioxidants produces a more pronounced activity than the sum of the activities of the individual antioxidants when used separately. One involves the action of mixed free radical acceptors. The other involves the combined action of a free radical acceptor and a metal chelating agent.

Safety and control : In general the total concentration must not exceed 0.02% by weight based on the fat content of the food.

GENERALIZED SCHEME FOR THERMAL DECOMPOSITION OF LIPIDS

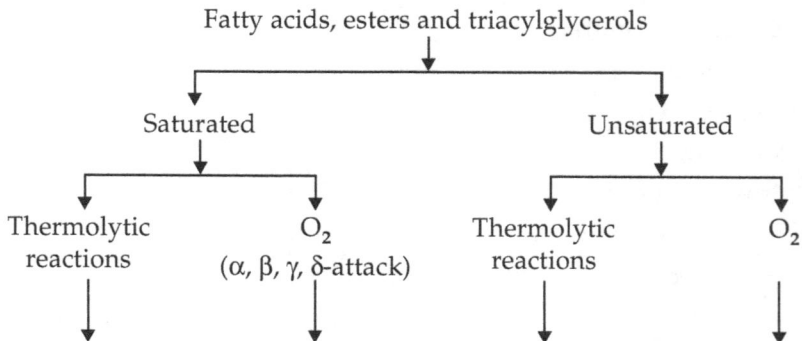

Fatty acids, esters and triacylglycerols

Saturated — Unsaturated

Thermolytic reactions — O_2 (α, β, γ, δ-attack) — Thermolytic reactions — O_2

Thermolytic reactions	O_2 (α, β, γ, δ-attack)	Thermolytic reactions	O_2
↓	↓	↓	↓
Acids hydrocarbons products propenediolesters and lactones of autoxidants acrolein ketones	long chain alkanes aldehydes, ketones	acyclic and cyclic dimmers	volatile and dimeric

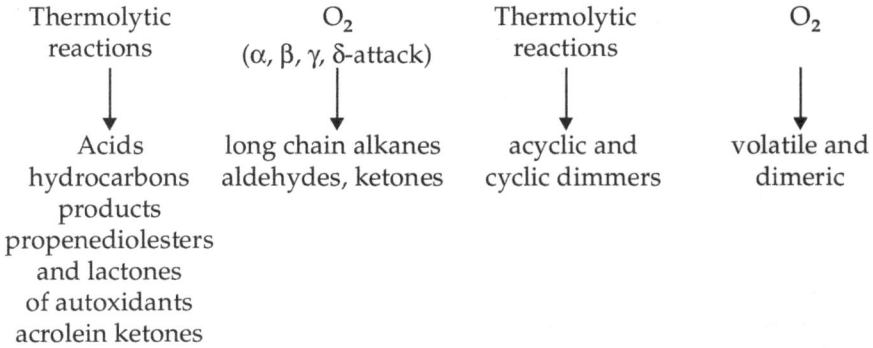

VOLATILES PRODUCED DURING FRYING

- Hexanal • Heptenal • Octenal • Decadienal (t, c) • Decadienal (t, t)
- Octane • Undecane • Pentylfuran • Pentadecane etc.

TESTS FOR ASSESSING THE QUALITY OF FRYING OILS

- Petroleum ether insoluble
- Column chromatography of polar compounds
- Dielectric constant
- Gas chromatography of dimmer esters

STEPS IN FAT AND OIL PROCESSING

- **Refining**
  - Settling and degumming
  - Neutralization
  - Bleaching
  - Deodorization
- **Hydrogenation**
  - Selectivity
  - Mechanism
  - Catalysts
    - Nickel
    - Copper
    - Chromium
    - Platinum
    - Palladium

- **Interesterification**
  - Principle
  - Industrial process
  - Mechanisms
  - Directed inter esterification
  - Applications

Important Sources of Plant Pils

Plant source	Oil (%)	Plant source	Oil (%)
Maize	5 - 10	Olive	15
Groundnut	40 - 45	Sesame	45 - 50
Soybean	15 - 45	Sunflower	30 - 40
Cottonseed	20 - 25	Oil palm	45
Linseed	35	Coconut flesh	65
Rapeseed	35 - 40	Castor bean	45 - 50

Lipid Content of Some Selected Food

Food	Lipid (%)	Food	Lipid (%)
Lard	99	Beef average	28
Cooking oil	100	Cream single	21
Margarine	85	Herring	14
Butter	83	Egg	12
Peanuts roasted	49	Chicken	6.7
Cream double	48	Milk	3.7
Chocolate milk	38	Cod	0.5

Fatty Acid Composition of Commonly Used Oils.

Oil	Saturated (%)	Monounsaturated (%)	Polyunsaturated (%)
Coconut	91	8	1
Cottonseed	34	26	40
Groundnut	20	54	26
Indian mustard	6	73	21
Niger	12	35	55
Palm	80	13	7
Safflower	11	13	76
Sesame	14	46	40

Oil	Saturated (%)	Monounsaturated (%)	Polyunsaturated (%)
Soybean	15	25	60
Sunflower	8	34	58
Rapeseed	4	16	14
Olive	11	76	7
Margarine	24	57	9
Almond	8	67	20
Cashew	17	70	7
Buffalo milk	44	26	1
Cow milk	37	33	3
Goat milk	35	25	5
Human milk	30	34	7

Analytical Data of Some Plant Oils.

Plant oil (15.5°C)	Sp. Gr. Ref. Index			
(40°C)	Iodine value	Saponification value		
Palm oil	0.858	1.451-1.459	49-57	196-202
Palm kernel oil	0.859-0.871	1.449-1.451	14-19	245-250
Olive oil	0.915-0.918	1.4605-1.4635	78-88	190-195
Coconut oil	0.869-0.874	1.448-1.450	7.5-9.5	255-260
Groundnut oil	0.917-0.919	1.4625-1.4645	85-100	100-196
Soybean oil	0.924-0.928	1.473-1.477	129-139	190-194
Cotton seed oil	0.920-0.925	1.4645-1.4655	103-115	190-198
Butter	-	-	25 - 28	230 - 240
Sunflower oil	-	-	125 - 135	-
Linseed oil	-	-	175 - 200	-
Human fat	-	-	-	195 - 200

Major Component Fatty Acids of Some Common Seed Oils.

Seed oil	Oil, %	Iodine value	$C_{16:0}$	$C_{18:0}$	$C_{18:1}$	$C_{18:2}$	$C_{18:3}$	$C_{22:1}$
Groundnut	45-51	85-100	6-14	2-7	40-72	13-38	-	-
Mustard	35-40	90-125	1-3	1-3	8-40	10-29	5-18	22-60
Safflower	35-40	130-150	2-10	1-6	7-42	55-80	0-3	-
Sesame	48-50	100-115	7-12	3-6	35-50	35-50	-	-
Soybean	13-20	115-140	7-12	2-6	20-50	36-65	2-13	-

Sunflower	40-42	115-145	3-10	1-10	14-72	20-75	-	-
Oil palm	50	53-60	44	4.5	37.2	10.1	0.4	-
Coconut	63	7.5-10.5	7-11	1-3	5-8	1-3	-	-
Cotton seed	15-23	95-115	17-29	1-4	13-44	33-58	-	-
Rice bran	15	-	18.1	1	41	38	1	-

Characteristics of Lipoproteins in Human Plasma

Characteristic	Chylomicrons	VLDL	LDL	HDL
Electrophoretic mobility	Origin	Pre-β	β	α
Density	0.96	0.96-1.006	1.006-1.063	1.063-1.21
Diameter (nm)	100-1000	30-90	20-25	10-20
Apoproteins	AI, AII, B_{48}	B_{100}, CI, CII, CIII, E	B_{100}	AI, AII, CI, CII, CIII, D.E.
Composition (%, approximate)				
Protein	2	10	20	40
Lipid (total)	98	90	80	60
Lipid components (%)				
Triacylglycerol	88	55	12	12
Cholesterol (free & ester)	4	24	59	40
Phospholipids	8	20	28	47
Free fatty acids	-	1	1	1

VLDL=Very low-density lipoproteins; LDL=Low-density lipoproteins; HDL=High-density lipoproteins.

What are the different detection tests of adulteration for edible oils?

1. Detection of sesame oil in edible oil (Boudouin Test):

   Oil/Fat + HCl, Furfural solution → Pink/red colour

2. Detection of cotton seed oil (Halphen's Test):

   Oil + Sulphur solution ("S" in carbon disulphide) + amyl alcohol Red colour

3. Detection of mineral oil (Holde's Test):

   Oil + Alcohol potassium hydroxide solution Turbidity.

4. Detection of castor oil:

   Castor oil + Separation on silica gel (TLC):

   Tririnolein + iodine vapours ⟶ Gives spot at Rf. 0.25.

5. Detection of argemone oil (cause: glaucoma, dropsy, blindness).

    Oil + HCl + Fe Cl$_3$ $\longrightarrow$ Needle like crystals.

6. Detection of hydrocyanic acid:

    Mostly mixed with mustard oil for good flavour.

    Oil + Tartaric acid $\longrightarrow$ 5% Na$_2$CO$_3$ solution, dipped picric acid paper + Stem $\longrightarrow$ Red colour.

7. Detection of fish/marine oils in edible oils.

    Oil + CHCl$_3$ (Wijs iodine solution, Br-HOAC) $\longrightarrow$ PPt formation

WHAT ARE MAIN CATEGORIES OF FAT REPLACERS?

The main categories of fat replacers are based on which component is attached with them, *i.e.,* carbohydrates, proteins or lipids.

Carbohydrate based:

*   Starch/modified food starch
*   Polydextrose
*   Inulin
*   Fiber
*   Nu-Trim

*   Dextrins   • Maltodextrins

*   Cellulose
*   Gums   • Polyols   • Z-trim
*   Oatrim

Protein based:

*   Microparticulated protein

*   Vegetable protein

*   Modified protein/denatured protein

*   Protein blends (contain carbohydrates and fat)

Lipid based:

*   Caprenin
*   Salatrim/Benefat • Captex
*   Bohenin
*   Esterified propoxylated glycerol
*   Trialkoxytricarballylate
*   Polyglycerol ester
*   Emulsifiers (mono- and diglycerides)
*   Fat blends (contain carbohydrates and fat)

*   Olestra/Olean
*   Neobee
*   Sorbestrin

*   Dialkyl dihexadecylmalonate
*   partially esteriifed polysaccharides

Chemical Composition of Normal Man (weight 65 kg)

Constituent	Percent	Weight (kg)
Water	61.6	40
Protein	17.0	11
Lipid	13.8	9
Carbohydrate	1.5	1
Minerals	6.1	4

WHAT ARE THE BIOLOGICAL FUNCTIONS OF LIPIDS?

- Lipids act as a prime fuel reserve for metabolism. It is concentrated source of energy, yielding more than twice energy (2.25 times) supplied by carbohydrates or proteins per unit weight.

- Fats help to reduce the bulk of the diet as starchy food.

- Fats are essential for the absorption of fat-soluble vitamins viz., A, D, E, K and especially carotenoids (vitamin A precursor) present in foods of vegetable origin.

- Fats improve the palatability of the diet.

- Some animal fat, e.g., fish liver oil, butter and ghee contain vitamin A and many vegetable foods contain vitamin E (wheat germ oil). Red palm oil is a good source of vitamin A.

- Fats are deposited in the adipose tissue. This deposit serves a reserve source of energy during starvation.

- Fat contains essential fatty acids viz., linoleic, linolenic and arachidonic fatty acids, which are essential for maintaining the tissues in normal health.

- Fats are essential for the utilization of galactose present in glucose.

- Nearly 25-30% of the dry weight of brain consists of phospholipids. Lipids are essential constituents of nervous tissue.

- Fat is often deposited, largely subcutaneous in warm-blooded animals and serves as insulation against in unfavorable environment, e.g., cold and physical injury.

- Waxes serves as a protective covering on the surface of an organism. The surface leaves, stems and fruits are rendered resistant to water, insects and bacteria by a wax coating.

- Fats are commercially important in the form of soaps, detergents grasses and the various oils of the paint industry.

- They are components of membrane (glycerophospholipids and sphingolipids).

  Several proteins are covalently modified by fatty acids.

- They act as energy store (triacylglycerols) and fuel molecules.

- Fatty acid derivatives serve as hormones and intracellular second messengers.

WHAT ARE KETONE BODIES?

When in excess acetyl CoA produced from the oxidation of fatty acid is converted into aceto acetate and D-3-hydroxybutyrate. Together with acetone these compounds are collectively termed ketone bodies. Aceto acetate and D-3-hydroxybutyrate are produced in the liver and provide an alternative supply of fuel for the brain under starvation conditions or in diabetes.

WHICH ARE THE STEPS INVOLVED IN PREPARATION OF RICE BRAIN OIL?

Rice bran contains 15% oil and is a very valuable source of oil. Rice brain oil is rich in unsaturated fatty acids particularly oleic and lenoleic acids. Steps involved in preparation of rice brain oil.

Steps	Purpose
Dewaxing	To remove wax
Degumming	To remove phospholipids
Neutralization or deacidification	To remove the fatty acids
Bleaching	To remove colour
Deodorization	To remove smell
Winterization glycerides	To remove saturated

WHAT IS WINTERIZING?

Triglycerides in vegetable oils are mixture of some long chain and more long-chain unsaturated fatty acids. At room temperature all fatty acids melt and remain liquid. But at refrigerated condition monounsaturated fatty acids become solid. Separation of liquid and solid fatty acids from lipids at cold condition is known as winterization. Example olive oil contains 75% monounsaturated fatty acids. Oil containing monounsaturated fatty acids becomes cloudy and viscous when refrigerated (cold storage). Cooling and filtration process is known as the winterization of lipids.

WHAT IS HOMOGENIZATION?

In the homogenization fat globules are broken up mechanically to less than 1 micron in diameter, so that fat does not rise to surface to form a cream layer. In homogenized milk the process consists of forcing milk heated to about 57 to 60°C through a very small orifice at high pressure. Infants more readily digest the fat in homogenized milk than that from ordinary milk.

WHAT IS EMULSIFICATION?

Emulsification is one of the properties of the fat that fats mixes with the water. It is one of the important properties of fats that influence the role of fat in cookery. Butter is an emulsion, so also is cream. Emulsions used as surface-active agents. Emulsifying power of surface-active agents is usually expressed by the so-called hydrophilic-lipophilic balance (HLB) system. Low HLB level to form water in oil (W/O) emulsion. High HLB level to form oil in water (O/W) emulsion.

AMINO ACIDS AND PROTEINS

Amino acids : Amino acids are organic compounds that contain amino and carboxyl group on the carbon atom and posses both acidic and basic properties.

Sr.No.	Group	Name	Abbreviation	Mol. Wt.
1.	Aliphatic	Glycine	Gly	75
		Alanine	Ala	89
		Valine	Val	117
		Leucine	Leu	131
		Isoleucine	Ile	131
2.	Hydroxylic	Serine	Ser	105
		Threonine	Thr	119
		Hydroxyproline	Hyp	131
3.	Sulphur	Cysteine	Cys	121
		Cystine	Cys-Cys	240
		Methionine	Met	149
4.	Acidic	Aspartic acid	Asp	133
		Glutamic acid	Glu	147
5.	Basic	Lysine	Lys	146
		Arginine	Arg	174
		Asparagine	Asn	132
		Glutamine	Gln	146

Sr.No.	Group	Name	Abbreviation	Mol. Wt.
6.	Aromatic (Hetrocyclic)	Phenylalanine	Phe	165
		Tyrosine	Tyr	181
		Trptophan	Try	204
		Histidine	His	155
7.	Imino acid	Proline	Pro	115

What are the essential and non-essential amino acids to the human being?

Sr. No.	Essential amino acids	Non-essential amino acids
1.	Valine	Glycine
2.	Leucine	Alanine
3.	Isoleucine	Serine
4.	Phenylalanine	Aspartic acid
5.	Tyrosine	Glutamic acid
6.	Tryptophan	Proline
7.	Lysine	Cysteine
8.	Methionine	Threonine
9.	Arginine (For adults)	Asparagine
10.	Histidine (For children)	Glutamine
11.		Hydroxyproline
12.		Hydroxylysine

- Those amino acids are not synthesized in human body are known as the essential amino acids.

- While protein synthesis in the plant or in human body those amino acids excost first are known as limiting amino acids.

- The pH at which amino acid or protein posses zero net charge and no migration in an electric field is known as the isoelectric point.

- Amino acids exist in neutral solution as doubly charged ions known as zwitterions and not as unionized molecules.

- **Protein :** It is macromolecule composed of one or more polypeptide chains, each posing a characteristic amino acid sequence and specific molecular weight. Protein also defined as a polymer of amino acids.

Classification of Proteins

- **Functional Classification**
  - Storage proteins
  - Transport proteins

- Catalytic proteins
- Contractile or motile proteins
- Structural proteins
- Defense proteins
- Immune proteins
- Regulatory proteins
- Other proteins
- **According to shape**
  - Globular proteins
  - Fibrous proteins
- **According to physical and chemical proteins**
  - Simple proteins
  - Conjugated proteins
  - Derived proteins
- **Combined proteins**
  - Membrane proteins
  - Plasma proteins
  - Hormonal proteins
  - Enzyme proteins
- **Structural proteins**
  - Primary structural proteins
  - Secondary structural proteins
    - α-Helix proteins
    - β-Plated sheet proteins
  - Tertiary structural proteins
  - Quaternary structural proteins

Simple proteins	*Complex proteins*
Protamines and Histones	Metal proteins
Albumins	Phosphoproteins
Globulins	Chromoproteins
Glutelins	Glucoproteins
Prolamins	Lipoproteins
Scleroprteins	Nucleoproteins
	Viruses

- **Seed Proteins :** Seed proteins from cereal grains are very heterogeneous and complex mixture of a large number of completely different types of molecules. All the individual molecular species of these proteins have not yet been isolated. Based on the solubility in various solvents, cereal proteins have been classified and separated into 4 major groups.

- **Albumin :** Water-soluble

- **Globulin :** Salt-soluble

- **Prolamine :** Alcohol-soluble

- **Glutelin :** Soluble in dilute alkali or acid

Albumin and globulin mainly function as enzymatic proteins, which carry out the metabolic processes in the grain, while prolamines, the major storage proteins are tissue-specific and are synthesized in the endosperm. They act as storage of nitrogen, carbon and generally sulphur and provide nutrition to the germinating embryo. Prolamines account for as much as 30-60% of the total grain protein in almost all cereals except rice. The essential amino acid composition of different protein fractions along with FAO reference protein indicates that albumin, globulin and glutelin are nutritionally more balanced while prolamines are extremely deficient in lysine and tryptophan. Since prolamines are the major proteins, they tend to lower the nutritional quality of cereal grains.

Protein Content and Prolamine Fraction of Cereal Grains

Cereal grain	Protein (% of dry weight)	Prolamine (% of total protein)
Wheat	10 -15	40 - 50 (Gliadin)
Maize	7 - 13	50 - 55 (Zein)
Barley	10 - 16	35 - 45 (Hordein)
Sorghum	9 - 13	60 (Kafirin)
Rice	8 - 10	1 - 5 (Oryzin)
Pearl millet	8 - 16	21 - 38 (Prolamine)

What are the classes of secretary protein in vertebrates?

Sr.No.	Type	Example	Sites of synthesis
1.	Serum proteins	Albumin	Liver (hepatocyte)
		Transferrin (Fe transporter)	Liver
		Immunoglobulins	Lymphocytes
		Lipoproteins	Liver
2.	Structural proteins	Collagen	Fibroblasts
		Fibronectin	Fibroblasts

Sr.No.	Type	Example	Sites of synthesis
3.	Peptide hormones	Insulin	Pancreatic β-islet cells
		Glucagons	Pancreatic α-islet cells
		Endorphins	Neurosecretory cells
		Enkephalins	Neurosecretory cells
		ACTH	Pituitary anterior lobe
4.	Digestive enzymes	Trypsin	Pancreatic acini
		Chymotrypsin	Pancreatic acini
		Amylase	Pancreatic acini, liver, Salivary glands
		Ribonuclease	Pancreatic acini
		Deoxyribonuclease	Pancreatic acini
5.	Milk proteins	Casein	Mammary gland
		Lactalbumin	Mammary gland
6.	Egg white proteins	Ovalbumin	Tubular gland cells of the avian oviduct
		Conalbumin	Tubular gland cells of the avian oviduct
		Lysozyme	Tubular gland cells of the avian oviduct

Present Sources of Energy (averaged over the earth's surface)

Sr.No.	Source	Energy in cal/(cm². Yr)
1.	Total radiation from sun	
2.	Ultraviolet light with wave length of	
	300 - 400 nm	$3.4 \times 10_3$
	250 - 300 nm	$5.6 \times 10_2$
	200 - 250 nm	$4.1 \times 10_1$
	< 150 nm	1.7
3.	Electric discharges	4
4.	Shock waves	1.1
5.	Radioactivity (to 1.0 km depth)	8×10^{-1}
6.	Volcanoes	1.3×10^{-1}
7.	Cosmic rays	1.5×10^{-3}

WHAT ARE THE QUALITATIVE TESTS FOR AMINO ACIDS AND PROTEINS?

The list of tests for amino acids and proteins testing is given below:

- Solubility test
- The ninhydrin reaction test
- The xanthoproteic reaction test
- Millon's reaction test
- Glyoxylic reaction test
- Pauly's test
- Ehrlich's reagent test
- The nitroprusside test
- The sakaguchi reaction test
- The biuret test
- C-terminal amino test
- N-terminal amino test
- The Folin-Lowery method test

WHAT ARE THE PROTEIN QUALITY PARAMETERS?

- Total protein content
- True protein content
- In-Vitro protein digestibility
- Net protein ratio
- True digestibility
- Protein efficiency ratio
- Biological value
- Calorific value
- Chemical score

WHAT ARE THE STEPS IN PROTEIN SYNTHESIS?

- Activation of amino acids
- Initiation of the polypeptide chain
- Elongation

- Termination
- Folding and processing
- What are the multification factors while calculating protein content from different food and food products?

  The factor 6.25 has been derived on the assumption that all the proteins content 16% nitrogen (100 ÷ 16 = 6.25). However, with advancement in techniques, several natural proteins have now been isolated and their nitrogen content has been accurately determined. It is seen that many proteins contain more than 16% nitrogen; hence factor for such protein samples is less than 6.25.

Sr.No.	Food material	Multification factor	Nitrogen content
1.	Sorghum, Bajara/pearl millet, other millets and maize	6.25	16.00
2.	Rice	5.95	16.80
3.	Whole wheat	5.83	17.15
4.	Wheat bran	6.31	15.85
5.	Wheat flour (free from germ or bran)	5.70	17.54
6.	Bulgur wheat	5.83	17.15
7.	Rye	5.83	17.15
8.	Barley, Oats	5.83	17.15
9.	Groundnut (peanuts) and Brazil nuts	5.46	18.31
10.	Soybean	5.71	17.51
11.	Sesamum	5.30	18.87
12.	Sunflower	5.36	18.66
13.	Almonds	5.18	19.30
14.	Other nuts and oilseeds	5.30	18.87
15.	Casein and other milk proteins	6.38	15.67

WHAT ARE THE DIFFERENT STEPS INVOLVED IN THE PROTEIN DETERMINATIONS?

1. *Oxidation (Digestion)*

   $$\text{Organic Nitrogen} + H_2SO_4 \longrightarrow (NH_4)_2SO_4$$

2. *Distillation*

   $$(NH_4)_2SO_4 + 2NaOH \longrightarrow Na_2SO_4 + NH_3 + 2H_2O$$

   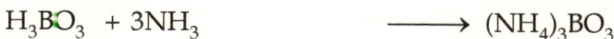

   $$H_3BO_3 + 3NH_3 \longrightarrow (NH_4)_3BO_3$$

3. *Titration*

   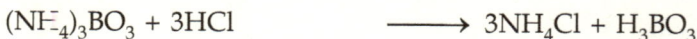

   $$(NH_4)_3BO_3 + 3HCl \longrightarrow 3NH_4Cl + H_3BO_3$$

One mole of nitrogen is equivalent to one mole of HCl

That is 36.5 g of HCl = 14 g of Nitrogen

1 N, 1000 ml of HCl = 14 g Nitrogen

1 N, 1 ml HCl = 0.014 g Nitrogen

WHAT IS MEAN BY "CENTRAL DOGMA"?

Protein Synthesis : The three roles of RNA. Describing the process of nucleic acid and protein synthesis in cells is a kin to describing a circle. The intricate relation between DNA, RNA, and protein can be diagrammed as follows:

$$DNA \longrightarrow RNA \longrightarrow Protein$$

DNA directs the synthesis of RNA; RNA then directs the synthesis of protein and special proteins catalyze the synthesis of both RNA and DNA. This cyclic flow of information occurs in all cells and it has been referred to as the "Central dogma" of molecular biology.

WHAT ARE THE BIOCHEMICAL FUNCTIONS OF PROTEINS?

- **Enzymes :** metabolic catalysts, oxidation, synthesis etc.
- **Structural proteins :** Keratins-proteins of hair, hair, feathers, collagen-connective tissue fibers, muscle proteins, silk and chitin.
- **Respiratory proteins :** Hemoglobin, cytochromes, myoglobin, and hemocyanin.
- **Antibodies :** Proteins formed in response to antigens, protection against foreign proteins.
- **Hormones :** Regulatory of metabolism, e.g., insulin controls carbohydrate and fat metabolism.
- **Nucleoproteins :** Chromosomal proteins, control of hereditary transmission, ribosomal proteins, involved in protein synthesis.
- To replace the daily loss of body proteins.
- To provide amino acids for the formation of tissue during growth.
- To provide the amino acids necessary for the formation of enzymes, blood proteins, hormones are proteins in nature.
- To provide amino acids for the growth of fetus in pregnancy and for the production of milk during lactations.
- The structural proteins are fibrous proteins or chains proteins, keratin in skin, hair and nails, myosin in muscles.

- Proteins provide essential amino acids, which are not synthesized in the animal body.

- The blood and water balance of animal depends upon the osmotic pressure of plasma, serum albumin level.

WHICH NITROGENOUS COMPOUNDS OCCURRING IN PLANTS?

- Ammonium

- Nitrate

- Proteins and their building blocks

- Enzymes of protein metabolism

- Fixation of elementary N

- Nitrification and denitrification

- Metabolism of proteins

- Amines and uridines

- Purines and pyrimidines

- Alkaloids

IONIC STRENGTH (μ)

Solubility of proteins mostly depends on ionic strength of the solution.

Ionic strength $(\mu) = \frac{1}{2} \; \Sigma C_i Z_i^2$

Where 'C' = concentration of solution, Z = valence of each atom.

The ions of neutral salts at molarities of the order of 0.5 to 1.0 M may increase the solubility of proteins. This effect is called "salting in". In this solution decrease the electrostatic attraction. If the concentration of neutral salts is greater than 1.0 M the protein displays a decreased solubility, which may result in precipitation. This effect is called "salting out" effect results from the competition between the protein and the salt ions for the water molecules necessary for their respective salvations.

DENATURATION OF PROTEINS

Protein denaturation is any modification in conformation (secondary, tertiary or quaternary) not accompanied by the rupture of peptide bonds involved in primary structure.

FACTORS RESPONSIBLE FOR DENATURATION OF PROTEINS

- **Physical agents**
  - Heat
  - Cold
  - Mechanical treatments
  - Hydrostatic pressure
  - Irradiation
  - Interfaces
- **Chemical agents**
  - Acids and alkalis
  - Metals
  - Organic solvents
  - Aqueous solutions of organic compounds

ENERGETIC OF DENATURATION

Protein denaturation is often irreversible especially when a strong denaturating treatment is administered. Very few cases it is "reversible/renaturation".

FOOD ENZYMES (ENZYMES USED IN FOOD INDUSTRY)

- Pectic enzymes
- Amylases
- Cathepsins
- Calcium-Activated
- Neutral proteinase
- Milk proteinase
- Lipolytic enzymes
- Thiaminases
- Phytase
- Myosin ATPase
- Enzymic browning
- Lipoxygenases
- Peroxidases

- Ascorbic acid oxidase
- Antioxidant enzymes
- Flavour enzymes
- Pigment-degrading enzymes

FACTORS USEFUL IN CONTROLLING ENZYME ACTIVITY IN FOOD SYSTEM

- Temperature
- Water content and water activity
- Extremes of pH
- Chemical used in food system
- Alterations of substrates
- Alterations of products
- Preprocessing control

The concept of total organic solids (TOS) in enzyme preparations allows an estimate to be made of the percent crude enzyme added to a food.

TOS (%) = 100 – A – W – D

Where: A=Percentage ash, W=Percentage of water, D=Percentage of diluents in the extract or isolated enzyme.

WHAT ARE THE ADVANTAGES OF IMMOBILIZED ENZYMES?

Following are the various advantages of immobilized enzymes.

1. Recovery and reuse
2. Stability increased
3. Kinetic property enhanced
4. Product is enzyme free
5. Permit continuous operation
6. Cost is lower
7. Greater control of catalytic power is possible

CARBOHYDRATES

How carbohydrates are classified into different groups?

Carbohydrates are carbon compounds that contain hydrogen and oxygen in the ratio 2:1. Chemically carbohydrates are aldehyde or ketone derivatives of polyhydric alcohols; these are known as aldoses or ketoses. Hydrated carbons are also called carbohydrates.

CLASSIFICATION OF CARBOHYDRATES

A. Monosaccharides

- Glyceraldehyde
- Dihydroxyacetone
- Erythrose
- Threose
- Xyloketose
- Arabinose
- Xylose
- Ribose
- Lyxose

- Lyxoketose
- Ribulose
- Xylulose
- Fructose
- Sorbose
- Glucose
- Mannose
- Galactose
- Gulose
- Idose

- Talose
- Atrose
- Allose
- Glucoheptose
- Mannoheptose
- Sedoheptulose
- Glucooctase
- Mannooctase
- Mannononose
- Glucodecose

B. Disaccharides

I. Reducing sugars

- Maltose α-1, 4 Glucose + Glucose
- Lactose β-1, 4 Glucose + Galactose
- Cellobiose β-1, 4 Glucose + Glucose
- Gentiobiose β-1, 6 Glucose + Glucose
- Melibiose α-1, 4 Glucose + Galactose
- Isomaltose α-1, 6 Glucose + Glucose

II. Non-reducing sugars

- Sucrose α-1, β-2 Glucose + Fructose
- Trehalose α-1, β-2 Glucose + Glucose

C. Trisaccharides

I. Reducing sugars

- Mannotriose $\longrightarrow$ Galactose + Galactose + Glucose
- Robinose $\longrightarrow$ Galactose + Rhamnose + Rhamnose
- Rhamninose $\longrightarrow$ Galactose + Rhamnose + Rhamnose

II.   Non-Reducing sugars

- Raffinose $\longrightarrow$ Fructose + Glucose + Galactose
- Gentobiose$\longrightarrow$ Fructose + Glucose + Glucose
- Melezitose $\longrightarrow$ Glucose + Fructose + Glucose

D. Oligosaccharides

- Stachylose $\rightarrow$ Fructose + Glucose + Galactose + Galactose
- Verbascose $\rightarrow$ Fructose + Glucose + Galactose + Galactose + Galactose
- Ajugose $\rightarrow$ Fructose + Glucose + Galactose + Galactose + Galactose + Galactose

E. Polysaccharides

I.   Homopolysaccharides

- Starches
- Dextrins
- Glycogens
- Celluloses
- Inulin
- Chitin
- Dextrans

II.   Heteropolysaccharides

- Agar
- Mucilages
- Gumarabic
- Gumacacia
- Pectins
- Alginic acids

III.   Mucopolysaccharides

- Hyaluronic acid
- Heparin
- Chondroitin sulfates
- Sialic acids

WHICH SACCHARIDES PRESENT IN HONEY?

Following different saccharides are present in honey:

Maltose, Sucrose, Gentiobiose, Isokestose, Kojibiose, Maltulose, Erlose, Melezitose, Turanose, Nigerose, Panose, Isomaltotetrose, Isomaltose, trehalose, Maltotriose etc.

Relative Sweetness of Different Sugars

Sugar	Relative sweetness
Sucrose	100
Fructose	170
Glucose	70
Galactose	32
Maltose	30
Lactose	16
Lactulose	55
Saccharin	40000 - 50000

Sweetness of Polyols

Sugar polyol	Relative sweetness
Gucitol	50
Lactitol	40
Maltitol	68
Mannitol	40
Xylitol	90
Sorbitol	63
Galactitol	58

WHAT ARE DIFFERENT SUGAR ALCOHOLS PRESENT IN HIGHER PLANTS?

- **Pentitol :** Ribitol
- **Hexitols :** Sorbitol, D-Mannitol, Dulcitol, Allitol, L-Iditol, Polygalitol, Styracitol, and Hamamalitol
- **Heptitols :** Volemitol, Perseitol
- **Octitol :** D-Erythro-D-galacto-octitol

Properties of Amylose and Amylopectin

Property	Amylose	Amylopectin
Molecular configuration	Essentially linear	Highly branched
Weight-average (mol. Wt.)	Ca. 10^6	Ca. 10^8
X-ray diffraction	Crystalline	Amorphous
Action of β-amylase and Z-enzyme	Complete hydrolysis	Residual dextrin of high mol.wt.
Complex formation	Readily forms complexes with iodine and polar substances	Very limited complexes formation
Stability in aqueous solution	Unstable, tends to retrograde	Stable

Starch Degrading Enzymes

Enzyme	Reaction catalyzed
α-amylase	Hydrolytic cleavage of α-1, 4 bonds in a glucan
β-amylase	Liberation of maltose from the non-reducing end of a glucan
α-glucosidase	Hydrolytic cleavage of α-1- 4, or α-1-6, bonds in oligosaccharides
Debranching enzyme	Hydrolysis of α-1-6 bonds
D-enzyme	G donor + G acceptor ↔ G donor + acceptor-1, + Glucose
Glucosyl-glucan transferase	Transfer of glycosyl, maltosyl or maltotriosyl residues to glucose
Phosphorylase	Glucose 1-P + α-glucan ↔ α-glucose 1-P + Glucose
Maltose phosphorylase	Matose + Orthophosphate ↔ Glucose-1-P + glucose

WHAT IS MEAN BY FIBER?

Dietary fiber includes all remnants of plant cells, which are resistant to digestion by the alimentary enzymes of humans.

Crude fiber (CF) = Cellulose

Acid detergent fiber (ADF) = Cellulose + Lignin

Neutral detergent fiber (NDF) = Cellulose + Lignin + Hemicelluloses

Total dietary fiber (TDF) = NDF + Pectin + Gum + Mucilage

Structural polysaccharides: Cellulose and noncellulosic polysaccharides

Structural nonpolysaccharides: Gums and mucilages

Plant cell wall ⟶ Cellulose

Structural ⟶ Polysaccharide ⟶ Hemicellulose

Non-carbohydrate polymer ⟶ Lignin

Plant non-structural substances and food additives pectin, gums, mucilages and modified polysaccharides.

- If there are four different atoms or functional groups singly bonded to a carbon atom in an organic molecule the carbon atom is said to be asymmetric? Since it can exist two isomeric forms called stereoisomers. The two different mirror images of a molecule, which are not super imposable, are called enantiomers.

- A molecule, which has only one asymmetric carbon atom that two ring, forms (α and β) and these are known as anomers.

- When D (+)-glucose is dissolved in water, a specific rotation of + 113 degrees is obtained, but this slowly changes, so that at 24 h the value has becomes + 52.5 degrees. This phenomenon is known as mutarotation.

- Monochromatic light passes through a Nicol prism and emerges polarized in one plane. This polarized beam then passes through the sugar sample, which rotates the plane of the light. This phenomenon is known as polarization or specific rotation of particular sugar.

Specific rotation $[\alpha]t = \alpha/l \times c$

Where: Path length (μ dm), concentration of solution (c, g/ml), Temperature (t, °C), Wave length (λ nm), $\lambda = 589$ nm, t = 20°C

- **Epimers :** Two stereoisomerisms differing in configuration at one asymmetric center in a compound having two or more asymmetric centers.

- Same molecular formula but different arrangement in the space of a molecule is known as isomerism.

Types of isomerism : 1. Structural isomerism and 2. Stereoisomerisms.

Structural isomerism : 1. Chain isomerism, 2. Positional isomerism and 3. functional group isomerism.

Stereoisomerism : 1. Optical isomerism, 2. Geometrical isomerism

- **The pasture effect :** When oxygen is admitted to an anaerobic suspension of cells utilizing glucose at a high rate, the rate of glucose consumption declined dramatically as the added oxygen is consumed. In addition the accumulation of lactate ceased. Louis Pasteur observed this effect in 1860 so it is known as Pasteur Effect.

- **Crabtree effect :** The Crabtree effect appears to be due to increased competition of glycolytic processes for inorganic phosphate and possibly pyridine nucleotides leaving less for oxidative phosphorylation reactions.

WHICH ARE THE MAJOR FACTORS INFLUENCES CARBOHYDRATE TOLERANCES?

Carbohydrate starvation

Effect of exercise

Hyper insulinism

Diseases of the liver

Acute and chronic infections

Thiamine deficiency

Nervous disorders

Effect of Anesthetics and Other Drugs

- Starch is storage homopolysaccharide mostly present in plants and it consists of amylose and amylopectin.

- The partial hydrolysis of starch by acids or α-and-β amylases produces small substances known as dextrins.

- The larger branched dextrins give a red colour with iodine and are called as "erythro-dextrin".

- The aldehyde and primary alcohol groups of aldoses can be oxidized to the corresponding aldonic or uronic acids and further oxidation of these acids yields Saccharic acid.

- Both aldoses and ketoses may be reduced to the corresponding polyhydroxy alcohols.

- The sugar alcohols are well-crystallized compounds soluble in water and alcohol and they have a sweet taste.

- Enolization takes place due to the action of alkalis on sugars.

- When sugars react with phenyl hydrazine it forms osazone.

- The hydrolysis of sucrose and formation of the equimolar mixture of glucose and fructose is known as invert sugar.

General tests for chemical properties of carbohydrates

- Molisch's test
- The anthrone reaction
- Benedict's test
- Barfoed's test
- Iodine test
- Osazone formation test
- Seliwanoff's test (Resorcinol hydrochloric acid test)

- Phloroglucinol-hydrochloric acid test
- Naphtho resorcinol test
- Benzidine test
- Preparation of osazones
- Bial's test for pentoses
- Tests for sucrose

How Many ATPs are Evolved During Complete Oxidation of Glucose?

Reaction sequence	*ATP yield/Glucose*
• Glycolysis: Glucose to pyruvate (in the cytosol)	
Phosphorylation of glucose	-1
Phosphorylation of fructose 6-phosphate	-1
Dephosphorylation of 2 molecules of 1,3-DPG	+2
Dephosphorylation of 2 molecules of phosphoenol pyruvate 2 NADH are formed in the oxidation of 2 molecules of glyceraldehydes 3-phosphate	+2
• Conversion of pyruvate to acetyl CoA (inside mitochondria) 2 NADH are formed	
• Citric acid cycle (inside mitochondria) Formation of 2 molecules of guanosine triphosphate from 2 molecules of succinyl CoA	+2
6 NADH are formed in the oxidation of 2 molecules of isocitrate, -ketoglutarate and malate	
2FADH$_2$ are formed in the oxidation of 2 molecules of succinate	
• Oxidative phosphorylation (inside mitochondria)	
2NADH formed in the glycolysis; each yields 2ATP (not 3ATP each, because of the cost of the shuttle)	+4
2NADH formed in the oxidative decarboxylation of pyruvate; each yields 3ATP	+6
2FADH$_2$ formed in the citric acid cycle; each yields 2ATP	+4
6NADH formed in the citric acid cycle; each yields 3ATP	+18
Net	**+36**

The overall reaction is : Glucose + 36ADP +36Pi +36H$^+$ + 6O$_2$ 6CO$_2$ +36ATP + 42H$_2$O

The P:O ratio is 3 since 36 ATP are formed and 12 atoms of oxygen are consumed. The vast majority of the ATP, 32 out of 36 is generated by oxidative phosphorylation. The overall efficiency of ATP generation is high. The oxidation of glucose yields 686 Kcal under standard conditions;

Glucose + 6O$_2$ 6CO$_2$ + 6H$_2$O ΔG0′ = −686 Kcal

The free energy stored in 36 ATP is 263 Kcal, since for hydrolysis of ATP is −7.3 Kcal. Hence, the thermodynamic efficiency of ATP formation from glucose is 263/686 or 38% under standard conditions.

WHAT ARE THE BIOCHEMICAL FUNCTIONS OF CARBOHYDRATES?

- Source of energy
- Conversion of solar energy in to chemical energy

- Central to metabolism
- Utilization of fats
- Sweetness of food
- Appetizing and flavouring
- Bulk of food
- Peristaltatic movement of stomach
- Movement of joints
- Structural support
- Storage of food material
- Transport between cells and organs
- Accumulation of fats
- Helps in pollination
- Helps in transamination
- Helps for fat burns
- Responsible for ketone body formation
- It exerts sparing effect of protein
- It acts as drugs
- It helps in protein synthesis
- Responsible for colour development to the products
- The ability to bind water and control water activity in foods is one of the most important properties of carbohydrates
- It helps in the sugar-flavourant formation
- A burnt carbohydrate gives colour to the food products
- It also acts as a lubricant at joints

The RDA for vitamin "A" is expressed as "retinol equivalents".

1 Retinol equivalent = 1 μ g retinol

= 6 μ g β-carotene

= 12 μ g other vitamin "A" active compounds.

- Classifications of vitamins:
    - Water-soluble/Energy releasing vitamins:
    - Ascorbic acid (Non-B-complex)
    - Thiamine (B_1)

- Riboflavin (B$_2$)
- Nicotinic acid (Niacin)
- Pyridoxine (Vitamin B$_6$)
- Biotin (B$_7$)
- Pantothenic acid
- Hematopoietic Vitamins
  - Folic acid
  - Cyanocobalamin (Vitamin B$_{12}$)
- Fat-soluble vitamins:
  - Vitamin A
  - Vitamin D
  - Vitamin E
  - Vitamin K

GENERAL CAUSES FOR LOSS OF VITAMINS AND MINERALS

- Genetics and maturity
- Handling postharvest or immediately postmortem
- Trimming
- Milling
- Leaching and blanching
- Processing chemicals
- Deteriorative reactions

Vitamins and Deficiency Symptoms

Vitamin	Deficiency
• Retinol (Vit. A)	Eye lesions (xerophthalmia, keratomalacia or blindness)
• Calciferol (Vit. K)	Bone malformations (rickets, osteomalacia)
• -Tocopherol (Vit. E)	Blood disorders (microcytic anemia and edema in prematures, creatinuria, red cell hemolysis)
• Phylloquinone (Vit. K)	Hemorrhage, decreased clotting of blood
• Thiamin (Vit.B$_1$)	Beriberi
• Riboflavin (Vit. B$_2$)	Mouth, skin and eye lesions
• Niacin (Vit. B$_3$)	Pellagra
• Folacin	Macrocytic anemia

Vitamin	Deficiency
• Biotin (Vit. H)	Seborrheic dermatitis
• Pantothenic acid	Gastrointestinal and nervous disorders
• Pyridoxine (Vit. B_6)	Convulsions, dermatitis
• Cobalamin (Vit. B_{12})	Pernicious anemia
• Ascorbic acid (Vit. C)	Scurvy
• Choline	Fatty liver

Physiological Functions of Minerals

Mineral (element in the form of a compound)	Physiological functions
• Calcium	Bone formation, various metabolic functions
• Phosphorus	Bone formation, various metabolic functions
• Magnesium	Bone constituent, carbohydrate and protein metabolism
• Sodium	Water and acid-base balance, pressure, muscle function
• Potassium	Acid-base balance, muscle function, carbohydrate and protein metabolism
• Chlorine	Acid-base balance, gastric digestion
• Sulphur	Various metabolic functions
• Iron	Hemoglobin synthesis, cellular oxidations
• Copper	Association with iron in enzyme systems and hemoglobin synthesis
• Iodine	Synthesis of thyroxin, the thyroid hormone
• Manganese	Activation of reactions in protein metabolism, glucose metabolism and fatty acid synthesis.
• Cobalt	Constituent of vitamin B, essential factor in red blood cell formation.
• Zinc	Essential enzyme constituent, involved in storage of insulin
• Molybdenum	Constituent of certain enzymes
• Fluorine	Dental health
• Selenium	Integrity of cell structures
• Chromium	Glucose metabolism

Vitamins: Roles and Clinical Signs of Deficiencies

Vitamin	Roles	Symptoms
• B_1 (Thiamin)	Coenzyme involved in the removal of CO_2 (-keto acids and glucose metabolism)	Beriberi, hypoglycemia, blood acidosis
• B_2 (Riboflavin)	Constituent of flavin nucleotide coenzyme related to energy metabolism	Photophosbia, glossitis, cheilosis, skin pruritus

Vitamin	Roles	Symptoms
• B$_6$ (Pyridoxine)	Coenzyme related to amino acid metabolism	Dermatitis, intertrigo, seborrhea, irritability, somnolence, neuropathy
• B$_{12}$ (Cobalamin) and Folic acid	Nucleic acid & amino acid metabolism	Megaloblastic anemia, glossitis, diarrhea, neuromyelopathy
• Ascorbic acid (Vit. C)	Maintenance of intercellular matrix of cartilage, bone & dentine, collagen synthesis, antioxidant	Scurvy
• Biotin	Coenzyme related to fat synthesis, amino acid metabolism & glycogen formation	Fatigue, depression, dermatitis, myalogia
• Niacin	Constituent of NAD & NADP related to reduction-oxidation reactions	Pellagra
• Pantothenic acid	Component of coenzyme-A related to energy metabolism	Asthenia, paresthesia, mental problem, epigastric discomfort.
• Choline	Constituent of phospholipids, precursor of neurotransmitter, acetyl choline	Not reported
• Vitamin A	Constituent of retinal pigment, maintenance of epithelium, infection defenses	Night blindness, xerothalmia, infection
• Vitamin D	Calcium absorption and deposition in bone	Osteomalacia, bone porosity, rckets
• Vitamin E & Selenium	Antioxidant, preventing cell membrane lesions	Myalgia, cardiomyopathy
• Vitamin K	Helps in blood clotting	Blood dyscrasia

Occurrences of Some Trace Minerals in Drinking Water and Foods.

Element	Concentration
• Arsenic	0-1000 µg/liter, 137-330 µg/day in food
• Barium	1 mg/liter
• Beryllium	1 µg/liter
• Cadmium	10 µg/liter
• Chromium	100 µg/liter
• Cobalt	Up to 0.5 mg/kg in green leafy vegetables
• Copper	1-280 µg/liter
• Lead	20-600 µg/liter, 100-300 µg per capita/day from food
• Manganese	0.5 - 1.5 mg/liter
• Magnesium	6 - 120 mg/liter
• Mercury	1 µg/liter, 10 µg/day intake from food
• Molybdenum	100 µg/liter, 100 - 1000 µg/kg of diet

Element	Concentration
• Nickel	1 - 100 µg/liter, 300 - 600 µg/day in food
• Selenium	10 µg/liter, 100 - 300 µg/kg in cereals, meats, sea foods
• Silver	Traces
• Tin	1 -2 µg/liter, 1 - 30 mg/day from food
• Vanadium	2 - 300 µg/liter
• Zinc	3 - 2000 µg/liter

FERMENTED MILK

Since milk goes bad very soon, it is converted into sour milk, which is acidic and hence will keep for longer time. Milk from different animals and different organisms are used for preparation of fermented milk in various countries. The consumption of soured-milk preparations is widespread because of their supposedly therapeutic value and also because they do not get spoilt as easily as milk. They appear under various names, which identify the country in which they are produced.

Curd : It forms an essential part of the diet in India and Ceylon. Curd is prepared from cow's or buffalo's milk, which is boiled, cooled and while still slightly warm, inoculated with a few drops from the previous day's curds. When allowed to ferment only for 5 or 6 hr, the curds are not sour, but within 10-12 hr, they become acidic. The organisms responsible for converting milk into curds are Streptococcus lactis and Lactobacillus.

Yoghurt : It is the sour milk preparation of countries like Bulgaria and Turkey. This is usually prepared from camel's or mare's though milk from other animals likes the cow are also sometimes used. Acid-forming organisms like Lactobacillus bulgaricus produce the fermentation; some times yeasts are also present and produce a small amount of alcohol and carbondioxide.

Kumiss : This is fermented milk prepared by the action of yeast, lactobacilli and streptococci on mare's or cow's milk used in Russia. The yeast produces alcohol and carbon-dioxdie and the bacteria produce lactic acid.

Leben : It is the sour milk prepared in Egypt from cow's milk or goat's milk or buffalo's milk, by the action of lactic acid producing bacteria and yeasts. The bacteria hydrolyse the lactose to glucose and galactose; some of the sugar is then fermented by the yeasts with the production of alcohol and carbondioxide and some converted into lactic acid by bacteria.

Kefir : It is prepared by the addition of kefir grains to milk. Kefir grains are cauliflower like aggregates of a mixture of microorganisms chiefly Saccharomyces, Lactobacillus casei and Streptococcus lactis and S. cremoris. The yeasts produce alcohol and carbondioxide and the bacteria produce lactic acid.

Matzoon : It is the sound milk preparation of Armenia and is similar to yoghurt in flavour and microflora.

Gioddin : Gioddin is the fermented milk preparation on the island of Sardinia. It contains the same organisms as Bulgarian yoghurt and Armenian matzoon.

Taette : It is ropy buttermilk made by means of a ropy variety of Streptococcus lactis.

Skyr : Skyr is semisolid fermented milk in which chiefly, Streptococcus thermophilus and Lactobacillus bulgaricus have been active.

CHANGES DURING MEAT COOKING

Meat is cooked to destroy any harmful microorganisms present in it and to improve its palatability. The changes brought about in meat are change in colour, weight, volume, flavour, fatty tissue and physical changes in the structural proteins, changes in the connective tissues and muscle fibre.

Changes in colour : The change in colour during cooking is due to denaturation of myoglobin and is converted to brown or dull red globin haemo chromogen. At the same time, some of the myoglobin is oxidized to metamyoglobin, which has a brown colour. Iron is in the ferric state in metamyoglobin. Hence cooked meat assumes a dull red to brown colour.

Change in flavour : The odour and taste of cooked meat arise from water and fat-soluble substances present in raw meat and by the liberation of volatiles formed during cooking. Water extract of raw meat develops a meaty flavour on heating. It is however weaker than the flavour of water in which meat is cooked. The volatiles of cooked meat contain H_2S, ammonia, acetaldehyde, acetone and diacetyl.

Change in juiciness : Juiciness depends on ability of meat proteins to retain the cooked water. The juiciness of cooked meat is greater at pH 4.0 or 7.0 than at pH 5.8. The damage in juiciness is correlated to shrinkage in cooking.

Shrinkage in volume and weight : The temperature of cooking affects both the rate and extent of shrinkage resulted in loss of volume. The pH of the meat seems to be most important factor in determining the loss in cooking. The shrinkage is greatest if the meat has a pH of 5.8, i.e., the isoelectric point of the main protein of muscle. The shrinkage is less at pH 4.0 or 7.0.

Change in fatty tissue : The changes in fatty tissue take place due to breakdown of collegenous tissue and escape of melted fat. The surface browning of fat may be due to oxidation of fat. The amount of fat found in drip varies depending on the fat content of meat.

Change in intracellular proteins and muscle fibres : The protoplasmic proteins found in muscle are denatured and become insoluble in water. The diameter and length of muscle fibre shrink.

Change in connective tissue : The connective tissue contains about 62% water, 32% collagen and 1.6% elastin. Collagen is converted into soluble gelatin during cooking of meat. The gelatin is gets dissolved in the water used for cooking. Elastin does not undergo any change during cooking.

Curing of meat : The additives salt or sugars owe their efficacy to their osmotic action, which deprives microorganisms, the available moisture. Initially there is an osmotic removal of water from the muscle proteins by 25-30% solution of sodium chloride. The final concentration of NaCl attained is 4-5%. Salt inactivates lipoxidase of muscle, because curing causes an acceleration of oxidative rancidity in the fat. In brines used for curing, 2.5 - 4% of potassium or sodium nitrate is included for its beneficial action on colour, although nitrate has a specific anti-microbial action especially in acid solution. A limit of 200 ppm is set by regulation for nitrites, as it is toxic at higher levels mainly through its effect on blood pigments. Salt improves water-binding capacity.

Smoking of meat : It is an additional process, which enhances meat preservation. Smoke contains phenols (derived from decomposition of lignin) which have anti-microbial action, specially enhanced by the drying action of the smoking process. Smoking also delays fat rancidity. The flavour of the smoked product depends partly on the reaction between the phenols and polyphenols with the SH groups in protein and between carbonyls and amino groups.

WHAT ARE INGREDIENTS PRESENT IN WOOD SMOKE?

1.	Formaldehyde	25-50 ppm
2.	Higher aldehydes	140-180 ppm
3.	Formic acid	90-125 ppm
4.	Acetic & higher ends	460-500 ppm
5.	Phenols	20-30 ppm
6.	Ketones	190-200 ppm
7.	Resins	1000 ppm

These compounds have antiseptic property. Generally smoking treatment to meat is given up to 18-24 hrs at 50-55°C.

FOOD ADDITIVES USED IN PROCESSED PRODUCTS

A. Introduction

A food additive is any substance that becomes part of food product either directly or indirectly during some phase of processing, storage, or packaging. There are two terms used in food science one is food additives and other is food ingredients. The difference between these two terms is mainly in the quantity used in any given formulation. Food ingredients can be consumed alone as food such as butter, sucrose etc., while food additives are mainly used in small quantities relative to the total food consumption but, which nonetheless play important role in the production of desirable and safe food products.

These food additives can be classified into two groups as direct food additives and indirect food additives. Direct food additives are those that have been intentionally added to food for functional purpose, in controlled amounts, usually at low levels mostly in ppm to 1-2 percent. Basic foodstuffs are excluded from the definition, although ingredients that are added to foods such as high fructose, corn syrup, starches, and protein concentrates are often included among food additives. Substances that come under the general definition of direct food additives give numbers in thousands and include both inorganic and organic chemicals, natural products, and modified natural and synthetic or artificial materials.

DIRECT FOOD ADDITIVES

- Inorganic chemicals: phosphates, sulfates, nitrates etc.
- Synthetic chemicals: dyes, silicones, benzoates, vitamin A, etc.
- Extraction products from natural sources: essential oils, gums, vitamin E, etc.
- Fermentation derived products: enzymes, lactic acid, citric acid, etc.

INDIRECT OR NON-INTENTIONAL ADDITIVES

Indirect or non-intentional additives are those entering into food products in small quantities as a result of growing, processing, or packaging.

The U.S. Code of Federal Regulations (CFR) provides classification for food additives as direct food additives into eight classes, secondary direct food additives into four classes and indirect food additives into eight classes based on their various categories.

DIRECT FOOD ADDITIVES AS PER THE CFR

- Food preservatives, e.g., sodium nitrate, sorbates etc.
- Coatings, films, and related substances, e.g., polyacrylamide
- Special dietary and nutritional additives, e.g., vitamins, trace minerals etc.
- Anticaking agents, e.g., sodium state, silicon dioxide.
- Flavouring agents and related substances, e.g., vanillin.

- Gums, chewing gum bases, and related substances, e.g., xanthan gum.
- Other specific usage additives, e.g., calcium lignosulfonate.
- Multipurpose additives, e.g., glycine.

SECONDARY DIRECT FOOD ADDITIVES AS PER THE CFR

- Polymer substances for food treatment, e.g., acrylate, acrylamide resins.
- Enzyme preparations and microorganisms, e.g., rennet, amylase.
- Solvents, lubricants, release agents, and related substances, e.g., hexane.
- Specific usage additives, e.g., boiler water additives, defoaming agents.

INDIRECT FOOD ADDITIVES AS PER THE CFR

- Components of adhesives, e.g., calcium ethyl acetoacetate 1, 4-butanediol modified with adipic acid.
- Components of coatings, e.g., acrylate ester copolymer coatings and polyvinyl fluoride resins.
- As components of paper and paperboard, e.g., slimicides, sodium nitrate/ urea complex, and alkyl ketene dimmers.
- As basic components of single and repeated use food contact surfaces, e.g., cellophane, ethylene-acrylic acid copolymers, isobutylene copolymers and nylon resins.
- As components of articles intended for repeated use, e.g., ultrafiltration membranes and textiles and textile fibers.
- Controlling growth of microorganisms, e.g., sanitizing solutions.
- Antioxidants and stabilizers, e.g., octyltin stabilizers in vinyl chloride plastics.
- Certain adjuvant and production aids, e.g., animal glue, hydrogenated castor oil, synthetic fatty alcohols and petrolatum.

The practice of adding desired chemicals such as salts, spices, herbs, vinegar and smoke to the food products dates back many centuries. But in the recent years, presence of chemical additives in processed foods has attracted much attention and public concern over the long term safety of additives to humans. There is scientific consensus that food additives are indispensable in the production, processing, and marketing of many food products, and that the judicious use of chemical additives typically in the range from a few ppm to <1% by weight of the finished food contributes to the abundance, variety, stability, microbiological safety, flavour, and appearance of the food supply. However, food additives offer a major contribution to the palatability and appeal of a wide variety of foods, their level of use is relatively insignificant in the total human diet.

B. VARIOUS FUNCTIONS OF DIFFERENT FOOD ADDITIONAL

Direct additives serve several major functions in food systems. Many food additives are acting as a multifunctional.

The basic functions are grouped in to five groups as follows:

- Extend shelf life, e.g., retard the onset of rancidity of oils.
- Ensure microbial safety, e.g., against botulism, listeria.
- Enhance acceptability and palatability, e.g., flavour, colour and texture.
- Improve nutritional value, e.g., vitamin and trace mineral fortification.
- Facilitate food processing, e.g., emulsifiers, anti-caking agents.

As per the FDA the group of food additives and their physical or technical effects are categorized as below:

S.No. Category of additive	Physical or technical effects (Functions)
1. Anticaking agent or free flow agent	Substance added to finely powdered or crystalline food product to prevent caking, lumping, or agglomeration
2. Antimicrobial agent	Substance used to preserve food by preventing growth of microorganisms and subsequent spoilage, including fungistats, mold and rope inhibitors. Also induces antimicrobial agents, antimyotic agents, preservatives and mold preventing agents.
3. Antioxidant	Substance used to preserve food by retarding deterioration, rancidity, or discoloration due to oxidation.
4. Boiler water additives	Substance used in a steam or boiler water system as an anticorrosion agent, to prevent scale or to effect steam purity.
5. Colour or coloring adjunct	Substance used to impart, preserve, or enhance the colour or shading of a food. Includes colour fixatives, colour-retention agents.
6. Curing or pickling agents	Substance imparting a unique flavor and/or colour to food, usually producing an increase in self life stability.
7. Dough strengthener	Substance used to modify starch and gluten, thereby producing more stable dough.
8. Drying agent	Substance with moisture absorbing ability, used to maintain an environment of low moisture.
9. Emulsifier or emulsifier salts	Substance which modifies surface tension in the component phase of an emulsion to establish a uniform dispersion or emulsion.

Contd...

S.No.	Category of additive	Physical or technical effects (Functions)
10.	Enzyme	Enzyme used to improve food processing and the quality of finished food.
11.	Firming agent	Substance added to precipitate residual pectin, thus strengthening the supporting tissue and preventing its collapse during processing.
12.	Flavour enhancer	Substance added to supplement, enhance or modify the original taste and/or aroma of a food, without imparting a characteristic taste or aroma of its own.
13.	Flavoring agent or adjuvant	Substance added to impart or help impart a taste or aroma in food.
14.	Flour treating agent	Substances added to milled flour, at the mill to improve its colour and/or baking qualities, including bleaching and maturing agents.
15.	Formulation aid	Substance used to promote or to produce a desired physical state or texture in food. Including carriers, binders, fillers, plasticizers, film formers and tableting aids
16.	Freezing or cooling agents	Substance that reduces the temperature of food materials through direct contact.
17.	Fumigant	Volatile substance used for controlling insects or pests.
18.	Humectants	Hygroscopic substances incorporating in food to promote retention of moisture. Includes moisture retention agents and antidusting agents.
19.	Leavening agent	Substance used to produce or stimulate production of carbon dioxide in baked goods in order to impart a light texture, including yeast, yeast foods, and calcium salts.
20.	Lubricant or release agent	Substance added to food contact surfaces to prevent ingredients and finished products from sticking to them. Includes release agents, lubricants, surface lubricants, waxes and anti-blocking agents.
21.	Malting or fermenting aids	Substance used to control the rate or nature of malting or fermenting process including microbial nutrients and suppressants and excluding acids and alkalis.
22.	Masticatory agents	Substance that is responsible for the long lasting and pliable property of chewing gum.
23.	Nonnutritive sweetener	Substance having <2% of the caloric value of sucrose per equivalent unit of sweetener capacity.
24.	Nutrient supplement	Substance necessary for the body's nutritional and metabolic process.
25.	Nutritive sweetener	Substance having >2% of sucrose per equivalent unit of sweetening capacity.

Contd...

S.No. Category of additive	Physical or technical effects (Functions)
26. Oxidizing or reducing agents	Substance, which chemically oxidizes or reduces another food ingredient, thereby producing a more stable product.
27. pH control agent	Substance added to change or to maintain active acidity or basicity. Includes buffers, acids, alkalis and neutralizing agents.
28. Processing aids	Substance used as a manufacturing aid to enhance the appeal or utility of a food or component. Includes clarifies, clouding agents, catalysts, flocculants, filter aids, crystallization inhibitors.
29. Propellant	Gas used to supply force to expel a product or used to reduce the amount of oxygen in contact with the food in packaging.
30. Sequesterant	Substance which combines with polyvalent metal ions to form a soluble metal complex to improve the quality and stability of products.
31. Solvent or vehicle	Substance used to extract or dissolve another substance.
32. Stabilizer or thickener	Substance used to produce viscous solutions or dispersions, to impart body, improve consistency or stabilize emulsions. Includes suspending and bodying agents, setting agents and bulking agents.
33. Surface active agent	Substance used to modify surface properties of liquid food components for a variety of effects, other than emulsifiers. Includes solubilizing agents, dispersants, detergents, wetting agents, rehydrating enhancers, foaming agents, defoaming agents.
34. Surface finishing agents	Substance used to increase palatability, preserve gloss and inhibit discoloration of foods. Includes glazes, polishes, waxes and protective coatings.
35. Synergist	Substance used to act or react with another food ingredient to produce a total effect different from or greater than the sum of the effects produced by the individual ingredients.
36. Texturizer	Substance, which affects the appearance or feel of the food.
37. Tracer	Substance added as a food constituent (as required by regulation) so that levels of this constituent can be detected after subsequent processing and/or combination with other materials.
38. Washing or surface removal agents	Substance used to wash or assist in the removal of unwanted surface layers from plant or animal tissues.

C. CLASSIFICATIONS OF VARIOUS FOOD ADDITIONAL

1. **Sweeteners polyols and bulking agents:**

   (a) **Sweeteners Polyols and Bulking Agents:** Sweeteners are used in formulated foods for many functional reasons as well as to impart sweetness. They render certain foods palatable and mask bitterness; and flavor, body, bulk, and texture; change the freezing point and control crystallization; control viscosity, which contributes to body and texture; and prevent spoilage. Certain sweeteners bind the moisture in food that is required by detrimental microorganisms.

   Sweeteners may be classified in a variety of ways:

   1. **Nutritive or Nonnutritive :** Materials either are metabolized and provide calories or are not metabolized and thus are noncalorie.

   2. **Natural or Synthetic :** Commercial products that are modifications of a natural product (e.g., honey or crystalline fructose is considered natural; saccharin is a synthetic compound).

   3. **Regular or Low Calorie/Dietetic/High Intensity :** Although two sweeteners may have the same number of calories per gram, one may be considered low calorie or high intensity if less material is used for equivalent sweetness.

   4. **As Foods :** For example, fruit juice concentrates can impart substantial sweetness.

   Polyols : Calorie Value and Relative Sweetness to Sugar.

S.No.	Polyol/ Sweetener	Chemical Formula	Relative sweetness, Sucrose=100	Calorie value, Kcal/g
1.	Erythritol	$C_4H_{10}C_4$	60-70	0.2
2.	Isomalt	-	45-65	2.0
3.	Lactitol	-	40	2.0
4.	Maltitol	$C_{12}H_{24}O_{11}$	90	3.0
5.	Mannitol	$C_6H_{14}O_6$	70	1.6
6.	Sorbitol	$C_6H_{14}O_6$	50-70	2.6
7.	Xylitol	$C_5H_{12}O_5$	100	2.4

   (b) **High intensity sweeteners:** High intensity sweeteners once used mainly for dietetic purpose, are now used as food additives in a wide variety of products. Termed "high intensity" because they are many-folds sweeter than sucrose and closely mimic its sweetness profile.

High Intensity Sweeteners.

S.No.	Sweetener	Chemical Formula	Sweetness, Sucrose = 1
1.	Acesulfame	$C_4H_5NO_4S.K$	200
2.	Alitame	$C_4H_7NO_4+C_3H_7NO_2$	2,000
3.	Aspartame	$C_{14}H_{18}N_2O_5$	200
4.	Cyclamate, Na Salt	$C_6H_{13}NO_3S. Na$	30
5.	Glycirrhizin	-	50
6.	Neohesperidin DC	-	2,000
7.	Neotame	-	7,000-13,000
8.	Saccharin	$C_7H_5NO_3S$	300
9.	Stevioside	-	300
10.	Sucralose	-	600
11.	Tagatose	-	-
12.	Thaumatin (Talin)	-	3,000
13.	Trehalose	-	-

(c) Bulking agents: Bulking agents are substances that add bulk to food products while contributing fewer calories than the ingredients they replace, e.g., polydextrose, specialty sweeteners etc.

2. FLAVOUR, TASTE, APPEARANCE AND TEXTURE ENHANCERS

(a) **Acidulants :** Acidulants complements fruit and other flavours in carbonated beverages, preserves, fruit drinks and desserts. Their ability to lower pH makes them useful as preservatives because an acid environment retards the growth of microorganisms responsible for spoilage and prevents enzymatic browning in fruit.

S.No.	Name of the compound	Chemical formula
Acidulants		
1.	Acetic acid	$C_2H_4O_2$
2.	Adipic acid	$C_6H_{10}O_4$
3.	Citric acid	$C_6H_8O_7$
4.	Fumaric acid	$C_4H_4O_4$
5.	Glucono-delta-lactone	$C_6H_{10}O_6$
6.	Lactic acid	$C_3H_6O_3$
7.	Malic acid	$C_4H_6O_5$
8.	Phosphoric acid	H_3PO_4
9.	Tartaric acid	$C_4H_6O_6$
10.	Vinegar	-

Flavours and Flavour Enhancers

1.	Essential oils and natural extracts	
2.	Aroma chemicals	
	(a) Anethole	$C_{10}H_{12}O$
	(b) Vanillin	$C_8H_8O_3$
	(c) Citronellol	$C_{10}H_{20}O$
	(d) Geraniol	$C_{10}H_{18}O$
	(e) Diacetyl	$C_4H_6O_2$
	(f) Benzaldehyde	C_7H_6O
3.	Falvour compositions	Moe than 100 constituents

Flavour constituents are added to foods and beverages in the following reasons:

- Create a totally new taste.
- Enhance, extended, round or increase the potency of flavours already present.
- Supplement or replace flavours to compensate for losses during processing.
- Simulate other more expensive flavours or replace unavailable flavours.
- Mask less desirable falvours to cover harsh or undesirable tastes naturally present in some food, other than hide spoilage.

Flavouring substances may be classified as:

- Natural Flavouring Substance
- Nature-Identical Flavouring Substance.
- Artificial Flavouring Substance.
- Flavouring Preparations.
- Process Flavourings.
- Smoke Flavouring.
- Flavour Enhancers.

(b) **Colours:** Colours are used in foods to improve appearance and thereby influence the perception of texture and taste.

Food colours may be added to food to:

- Give attractive appearance to foods that would otherwise look unattractive or unappetizing and thus enhance enjoyment.
- Restore the original appearance of the food where the natural colours have been destroyed by heat processing and with subsequent storage.

- Intensify colours naturally occurring in foods where the colour is weaker than that which the consumer associates with a food of that type of flavor.

- Ensure uniformity of colour due to natural variations in colour intensity.

- Help protect flavor and light-sensitive vitamins during shelf storage by a sunscreen effect.

- Help preserve the identity or character by which foods are recognized.

- Serve as visual indication of quality thus in addition to enhancing the acceptability of foods, colours, aid in food manufacture, storage and quality control.

Certified Colours Permitted for Use in Various Food Items.

S.No.	FDA name	Common name
1.	FD and C Blue No.1	Brilliant Blue
2.	FD and C Blue No. 2	Indigotine
3.	FD and C Green No. 3	Fast Green
4.	FD and C Yellow No. 5	Tartrazine
5.	FD and C Yellow No. 6	Sunset Yellow
6.	FD and C Red No. 3	Erythrosine
7.	FD and C Red No. 40	Allura Red
8.	Orange B	-
9.	Citrus Red No. 2	-

List of Exempted Colours from Certification

S.No.	Name	Colour	Source
1.	Annatto extract	Yellow	Vegetable
2.	Beet juice	Red	Vegetable
3.	Beets, dehydrated (powder)	Purple	Vegetable
4.	Canthaxanthin	Red	Synthetic
5.	Caramel	Brown	Semi Synthetic
6.	Apo-carotenal	Orange	Synthetic
7.	Beta-carotene	Yellow	Synthetic
8.	Carrot oil	Yellow	Vegetable
9.	Chlorophyll	Green	Vegetable
10.	Cochineal extract	Red	Insect
11.	Corn endosperm oil	Yellow	Vegetable

Contd...

S.No.	Name	Colour	Source
12.	Dried algae meal	Yellow	Plant
13.	Ferrous gluconate	Black	Synthetic
14.	Ferrous lactate	Black	Synthetic
15.	Fruit juice (grape, cranberry)	Red	Fruit
16.	Grape skin extract (enocyanin)	Red	Fruit
17.	Paprika	Red	Vegetable
18.	Paprika oleoresin	Red	Vegetable
19.	Red cabbage juice	Red	Vegetable
20.	Riboflavin	Yellow	Synthetic
21.	Saffron	Yellow	Vegetable
22.	Titanium dioxide	White	Synthetic
23.	Turmeric (curcumin)	Yellow	Vegetable
24.	Turmeric oleoresin	Yellow	Vegetable

(c) **Thickeners and Stabilizers :** Thickeners and stabilizers (also called hydrocolloids, gums or water-soluble polymers) provide a number of useful functions to food products. Most thickeners and stabilizers are polysaccharides and function in foods as:

- Rheology modifies, affecting the flow and feel (mouth) of food and beverage products.
- Suspension agents for food products containing particulate matter.
- Stabilizers of oil-water mixtures.
- Binders in dry and semidry food products.
- Gel-forming agents in food that require this physical form.

Two principal classes of these materials are utilized in food products.

(a) Natural materials obtained from plants or animals, including gum Arabic, locust bean gum, guar gum, alginates, carrageenan, pectin, starches, casein, gelatin etc.

(b) Semi synthetic materials, which are manufactured by chemical derivatization of natural organic materials or by microbial fermentation. This group includes carboxymethylcellulose (CMC) and other modified cellulose compounds, dextrin and gellan and xanthan gums.

(d) **Emulsifiers :** Emulsifiers or substances are additives that allow normally immiscible liquids, such as oil and water to form a stable mixture. Emulsifiers are widely used in foods in order to perform one or more of the following functions:

- Increase stability and prevent phase separation in food emulsions, e.g., mayonnaise, salad dressings.

- Improve the shelf life of flavours and related onset of rancidity in fats and oils containing food emulsions.

- Improve texture; reduce crumb firmness and complex with starches (baked goods).

List of Emulsifiers

Mono- and diglycerides (GRAS)	Succinyl monoglyceride
Lactylated monoglyceride	Acetylated monoglyceride
Monoglyceride citrate	Monoglyceride phosphate
Steryl- monoglyceride citrate	Diacetyl-tartrate ester of monoglyceride
Polyoxyethylene monoglyceride	Polyoxyethylene stearate
Propylene glycol monoester	Lactykated propylene glycol monoester
Sorbitan monosterate	Sorbitan tristearate
Polysorbate 60	Polysorbate 65
Polysorbate 80	Calcium stearoyl lactate
Sodium stearoyl lactylate	Stearoyl lactylic acid
Stearyl tartarate	Stearyl monoglyceridyl citrate
Sodium stearyl fumarate	Sodium lauryl sulfate
Dioctyl sodium sulfosuccinate	Polyglycerol esters
Sucrose esters	Sucrose glycerides
Lecithin (GRAS)	Hydroxylated lecithin
Triethyl citrate (GRAS)	

(e) **Bleaching, maturing and dough-conditioning agents :** Some chemicals serve as both bleaching and maturing agents and other are referred to as dough-conditioning agents or bread improvers, it is perhaps desirable to consider all of them under one heading. Bleaching agents are used in the production of certain cheeses, processed fruits, crude fats and oils and meat products to neutralize colour that may be present naturally.

List of Bleaching, Maturing and Dough-Conditioning Agents

S.No.	Name of the compound	Chemical formula
1.	Sulfuric acid	H_2SO_4
2.	Meta-phosphoric acid	H_3PO_4
3.	Hydrogen peroxide	H_2O_2
4.	Calcium hypochloride	$Ca(OCl)_2$
5.	Benzoyl peroxide	$C_{14}H_{10}O_4$

Contd...

S.No.	Name of the compound	Chemical formula
6.	Chlorine dioxide	ClO_2
7.	Nitrosyl chloride	$NOCl$
8.	Nitrogen oxides	N_2O
9.	Nitroen tetroxide	N_2O_4
10.	Potassium bromate	$KBrO_3$
11.	Potassium iodate	KIO_3
12.	Calcium iodate	$Ca(IO_3)_2$
13.	Calcium peroxide	CaO_2

(f) **Firming agents :** Fruits and vegetables contain pectin compounds that are relatively insoluble and form a firm gel around the fibrous tissues of the fruit and prevent its collapse. Addition of calcium salts causes the formation of calcium pectate gel, which supports the tissues and affords protection against softening during processing.

Some of the Examples of the Firming Agents

S.No.	Name of the compound	Chemical formula
1.	Calcium chloride	$CaCl_2$
2.	Calcium citrate	$Ca_3(C_6H_5O_7)_2$
3.	Monocalcium dihydrogen phosphate	$Ca(H_2PO_4)_2$
4.	Calcium lactate	$CaC_6H_{10}O_6$
5.	Calcium sulfate	$CaSO_4$
6.	Aluminum sulfate	$Al_2(SO_4)_3$
7.	Ammonium aluminum sulfate	$(NH_4)Al(SO_4)_2$
8.	Potassium aluminum sulfate	$KAl(SO_4)_2$
9.	Sodium aluminum sulfate	$NaAl(SO_4)_2$

(g) Glazing and polishing agents: These agents are used on coated confections to give luster to the otherwise dull coating. Chemicals that are used for this purpose include acetylated monoglycerides, beeswax, carnauba wax, gum Arabic, magnesium silicate, mineral oil, petroleum, shellac and zein.

3. PRESERVATIVES

Preservatives may be divided into two main groups:

Antimicrobial agents : Antimicrobial agents are capable of retarding or preventing growth of microorganisms such as yeast, bacteria, molds or fungi and subsequent spoilage of foods. The principal mechanisms are reduced water availability and increased acidity. Sometimes these additives also preserve other important food characteristics such as flavor, colour, texture and nutritional value.

Antioxidants : Antioxidants are food additives that retard atmospheric oxidation and its degrading effect, thus extending the shelf life of foods.

The Primary Food Additives used as Preservatives are as Follows:

S.No.	Name of the chemical	Chemical formula
1.	Benzoic acid	$C_7H_6O_2$
2.	Sodium benzoate	$C_7H_5O_2.Na$
3.	Potassium benzoate	$C_7H_5O_2.K$
4.	Sorbic acid	$C_6H_8O_2$
5.	Sorbic acid with sodium salt	$C_6H_8O_2.Na$
6.	Sorbic acid with potassium salt	$C_6H_8O_2.K$
7.	Propionic acid	$C_3H_6O_2$
8.	Propionic acid with calcium salt	$C_6H_{10}CaO_4$
9.	Propionic acid	$C_6H_7O_2.Na$
10.	Parabens: P-hydroxybenzoic acid methyl ester and	$C_8H_8O_3$
	P-hydroxypropyl benzoate	$C_{10}H_{12}O_3$
11.	Organic acids:	
	Acetic acid	$C_2H_4O_2$
	Citric acid	$C_6H_8O_7$
	Malic acid	$C_4H_6O_5$
	Lactic acid	$C_3H_6O_3$
	Adipic acid	$C_6H_{10}O_4$
	Tartaric acid	$C_4H_6O_6$
	Caprylic acid	$C_8H_{16}O_2$
12.	Suphfur dioxide	SO_2
	Potassium suphfate	$O_3S.2K$
	Potassium metabisulphate	O_5S_2K
	Sodium bisulfite	$HO_3S.Na$
	Sodium metabisulfite	$O_5S_2.2Na$
	Sodium sulphite	$O_3S.2Na$
13.	Nitrates and nitrites	NO_3 and NO_2
14.	Natural alternatives	
	Natamycin	$C_{33}H_{47}NO_{13}$
	Nisin	$C_{143}H_{230}N_{42}O_{37}S_7$

Food Antioxidants and Their Chemical Formula and Manufacturing Processes.

S.No.	Name of compound	Chemical formula	Manufacturing process
Oil soluble antioxidants			
1.	Butylated hydroxyanisole (BHA)	$C_{11}H_{16}O_2$	Synthesis
2.	Butylated hydroxytolune (BHT)	$C_{15}H_{24}O$	Synthesis
3.	Tert-butyl-hydroxyquinone (TBHQ)	$C_{14}H_{22}O_2$	Synthesis
4.	Propyl gallate (PG)	$C_{10}H_{12}O_5$	Synthesis
5.	Tocopherols	-	Extraction/ Synthesis
6.	Thiodipropionic acid	-	Synthesis
7.	Dilauryl thiodipropionate	-	Synthesis
8.	Ascorbyl palmitate	$C_{22}H_{38}O_7$	Synthesis
9.	Ethoxyquin	-	Synthesis
Water soluble antioxidants			
10.	Ascorbic acid	$C_6H_8O_6$	Fermentation
11.	Sodium ascorbate	$C_6H_7NaO_6$	Fermentation
12.	Erythorbic acid	$C_6H_8O_6$	Fermentation
13.	Sodium erythorbate	$C_6H_7NaO_6 \cdot H_2O$	Fermentation
14.	Glucose oxidase/catalase enzymes	-	Fermentation
15.	Gum guaiac	-	Extraction
16.	Sulfites	-	Synthesis
17.	Rosemary extract	-	Extraction

- **pH adjusting agents :** These agents also known as acids, alkalis, buffers and neutralizers. These chemicals are used in most segments of the food processing industries such as:

- The baking industry as chemical leavening agents.

- In soft drinks to provide tartness.

- In certain dairy products to adjust the acidity.

- In cheese spreads for emulsification.

- In confectionery products as flavouring, to control the degree of inversion of sugars and to control the texture in the processing of chocolates.

- In jams, jellies to provide proper gel formation.

Some solutions are used as buffers:

- **Phosphoric acid :** dibasic potassium phosphate.
- **Formic acid :** Sodium formate.
- **Acetic acid :** Sodium acetate.

- **Sodium bicarbonate :** Sodium carbonate.
- **Dibasic sodium phosphate :** Sodium hydroxide.

List of Some Important Chemicals with Formula

S.No.	Name of the chemical	Chemical formula
1.	Citric acid	$C_6H_8O_7$
2.	Phosphoric acid	H_3O_4P
3.	Acetic, adipic	$C_6H_{10}O_4$
4.	Fumaric	$C_4H_4O_4$
5.	Hydrochloric	HCl
6.	Lactic	$C_3H_6O_3$
7.	Malic	$C_4H_6O_5$
8.	Tartaric	$C_4H_6O_6$
9.	Glucono-delta-lactone	$C_6H_{10}O_6$
10.	Ammonium bicarbonate	$HCO_3 . H_4N$
11.	Ammonium hydroxide	$H_4N . HO$
12.	Calcium carbonate	$CO_3 . Ca$
13.	Calcium oxide	CaO
14.	Potassium bicarbonate	$KHCO_3$
15.	Potassium hydroxide	HKO
16.	Sodium bicarbonate	$NaHCO_3$
17.	Sodium carbonate	$CO_3 . 2Na$
18.	Sodium hydroxide	NaOH
19.	Sodium sequicarbonate	$Na_2CO_3NaHCO_3 . 2HO$

- **Fumigants for insect and pest control :** These compounds are volatile in nature such as ethylene oxide [C_2H_4O], Propylene oxide [C_3H_6O], methylene bromide [CH_3Br] etc.

- **Gases :** Gases provide three basic functions as food ingredients: preservation, carbonation, and aeration. For this purpose nitrogen [N_2], carbon dioxide [CO_2], nitrogen oxide [N_2O], ozone [O_3] gases are used.

- **Sequestering agents :** These agents also called chelates combine with polyvalent metal ions to form a soluble metal complex to improve the quality and stability of products as free metallic ions promote oxidation of food. This group includes ethylenediaminetetraacetic acid [EDTA], calcium acetate, calcium gluconate, calcium sulphate, citric acid, stearyl citrate, tartaric acid, sodium tartarate, calcium monoisopropyl citrate, sodium hexa-meta-phosphate, phosphoric acid, potassium citrate and various calcium, potassium and sodium phosphates.

4. PROCESSING AIDES

There are several chemicals that are generally used in food processing industry for various purposes.

(a) **Anticaking agents:** Dry food products that contain hygroscopic substances require the addition of an anticaking agent. These additives must be insoluble in water, have the capacity of absorbing excess moisture or by coating particles making them water repellent. These compounds include calcium silicate [$CaSiO_3$], calcium state [$C_{36}H_{70}CaO_4$], sodium silicoaluminate [Na_2SAlO_3], tri-calcium phosphate [$Ca_3(PO_4)_2$], magnesium silicate [$MgSiO_3$], magnesium carbonate [$MgCO_3$], sodium ferrocyanide [$Na_4Fe(CN)_6 . 10H_2O$] etc.

(b) **Antifoaming (defoaming) agents:** Foams caused by proteins or gasses are eliminated or suppressed by these agents such as oleic acid, silicon dioxide, white mineral oil and other fatty acids.

(c) **Enzymes:** Enzymes are used in the food industry for the specific purpose.

- Speed up reactions.
- Reduce viscosity.
- Improve extractions.
- Carry out bioconversions.
- Enhance separations.
- Develop functionality.
- Create/intensify flavour.
- Synthesize chemicals.

    Food enzymes usually classified into following various groups based on their functions:

- Carbohydrases, *Proteases, *Lipases, *Pectic enzymes, *Specialty enzymes etc.

Area of Food Enzymes Used and Their Functions.

S.No.	Area for enzyme used	Functions of enzymes
1.	Dairy products	Milk coagulation, milk protein modification, cheese flavour development, enzyme modified cheeses and removal of hydrogen peroxide
2.	Baking and cereals	Antistaling, dough improvement, improved crust colour, gluten hydrolysis
3.	Sugar processing	Removal of starches and processing from cane sugar
4.	Starch processing	Starch modification, liquefaction, isomerization, saccharification, modification and increasing yield

Contd...

Contd...

S.No. Area for enzyme used	Functions of enzymes
5. Oils and fats	Improving yield, inter-esterification, oil extraction and lecithin production
6. Flavours	Synthesis of flavour, production of natural esters
7. Alcohol fermentation	Starch liquification, improving yeast growth
8. Brewing	Adjunct liquification, enhanced fermentability, filtration, improvement, production of light beer and removal of protein haze
9. Fruit juices, wines	Increasing press yields, juice clarification, shelf life extension
10. Coffee processing	Separation of bean, viscosity control of concentrate
11. Chemical processes	Biotrasformation, synthesis
12. Analytical	Tests for dietary fiber, sugars
13. Waste treatment	Breakdown of cellulose, lignin, oil residues and other solid waste material

(d) **Humectants:** In certain foods, it is necessary to control the amount of water enters or exist the product. For this purpose polyols are generally used such as propylene glycol $[C_3H_8O_2]$, glycerol $[C_3H_8O_3]$, sorbitol $[C_6H_{14}O_6]$, mannitol $[C_6H_{14}O_6]$ etc.

(e) **Leavening agents:** Many bakery products, such as self-rising flours, prepared baking mixes and refrigerated dough, rely on chemical leavening agents to produce the gas that gives them volume. These agents include sodium bicarbonate $[NaHCO_3]$, ammonium bicarbonate $[NH_4HCO_3]$, potassium bicarbonate $[KHCO_3]$, potassium acid tartrate $[K_2C_4H_4O_6]$, sodium aluminum sulfate $[AlNa_9(SO_4)_2]$, glucono-delta-lactone $[C_6H_{10}O_6]$ etc.

(f) **Lubricants and release agents:** These agents are substances added to food processing equipment to prevent food ingredients and finished products from sticking to them. These agents include calcium state, magnesium carbonate, magnesium silicate, magnesium stearate, mannitol, mineral oil, mono- and diglyvlerides, sorbitol, stearic acid, lecithin and starch.

(g) **Manufacturing aids:** These aids including catalysts filter aids, clarifying and clouding agents are used to improve the appearance or performance of food products.

(h) **Solvents:** Solvents are generally used to either extract particular compounds such as an essential oil from a plant or to carry additives into a food system such as a flavour into a powdered mix.

(i) **Water-correcting agents:** Water used in the beverage industries is often corrected to uniform mineral salt content that corresponds to water known to give the most satisfactory final product. These agents include monoammonium phosphate [$NH_4H_2PO_4$], diammonium phosphate [$H_6N_2 \cdot H_3O_4P$], calcium dihydrogen phosphate [$Ca(H_2PO_4)_2$], calcium chloride [$CaCl_2$], salt sulphates, potassium chloride [KCl].

(j) **Nutrients:** Those nutrients are deficient in the food product are generally added as a supplement or nutrient enrichment and make that food good, nutritious to all people. Generally for this purpose dietary fibers and vitamins [vit. A, B-complex, D] and some selected minerals [Fe, I] are added as per the government rule and regulations.

D. GOVERNMENT REGULATIONS

The application of food additives is highly regulated worldwide although regulatory philosophy, the approval of specific product, and the level of enforcement differ from country to country. The United States, Western Europe and Japan are major industrial regions and these are the largest consumers of food additives. With only 13% of the world's population these countries account over two-third of the food additive market. In the United States, FDA, USDA, MID, BATF and FD&C are the various agencies regulate the use of food additives in various food products in the country. Food regulations are based on the 1958 Food Additives Amendment to the Food, Drug and Cosmetic (FD&C) Act of 1938. The amendment was enacted with the following purpose.

- Protecting public health by requiring proof of safety before a substance could be added to food.
- Advancing food technology.
- Improving the food supply by permitting the use of substances in food that is safe at the levels of intended use.

According to the legal definition, food additives that are subject to the amendment include "any substance the intended use of which results or may reasonably be expected to result directly or indirectly in its becoming a component or otherwise affecting the characteristics of any food". This definition includes any substance used in the production, processing, treatment, packaging, transport or storage of food. If a substance is added to a food for a specific purpose it is referred to as a direct additive, e.g., sweetener aspartame. As per the regulations indirect food additives are those that become part of the food in trace amounts due to its packaging, storage or other handling.

FOR REGULATORY PURPOSE ALL FOOD ADDITIVES FALL INTO ONE OF THREE CATEGORIES GIVEN BELOW

- Generally Recognized As Safe (GRAS) substances.
- Prior sanctioned substances.
- Regulated-direct-indirect additives.

GRAS substances (~ 700) are a group of additives regarded by qualified experts as "generally recognized as safe". These substances are considered safe because their past extensive use has not shown any harmful effects.

Prior sanctioned substances (~ 1400) are products that already were in use in foods prior to the 1958 Food Additives Amendment to the federal Food, Drug and Cosmetic Act, and are therefore considered exempt from the approval process.

All other additives are regulated i.e., a specific food additive petition must be filed with the FDA, requesting approval for use of the additive in any application not previously approved.

All colour additives are subject to the Colour Additive Amendment of 1960. Colours permitted for use in foods are classified either as certified or exempt from certification. Colour additives that are exempt from certification include pigments derived from natural sources. However, colour exempt from certification also must meet certain legal criteria for specifications and purity.

Flavour substances are regulated somewhat differently, and the rules are less restrictive. However, the use of aroma chemicals as flavour ingredients is regulated also under laws that may differ from country to country.

Under FDA, USDA, and BATF regulations, the ingredients of a food or beverage must be stated on the product label in decreasing order of predominance. For many direct additive categories, chemical constituents must be identified by their common names and the purpose for which they were added. The FDAs Food Additives Amendment also contains what is known as the Delaney Clause, which mandates, regardless of dose level or intended use. The Delaney Clause is totally inflexible in that it does not recognize any threshold level below which the additives might not present a health hazard. Thus, it has caused a number of problems for the food industry and for food additives.

Approval Process: A new substance gains approval for food use through the successful submission of a Food Additive petition that must document the following points:

1. Safety, including chronic feeding studies in two species of animals.

2. Intended use

3. Efficiency data at specific levels in specified food systems.

4. Manufacturing details and product specifications.

5. Methods for analysis of the substance in food.

6. Environmental impact statement.

The other pathways are a somewhat simpler route. This is the GRAS affirmation process by which a petitioner can affirm that a substance either has a long history of use in the food supply and/or is generally recognized as safe (GRAS) by experts in the field.

In the European Union (EU) formation of the European Economic Community has created a requirement to bring food additive approvals of the member nations into alignment, so as to eliminated differences in laws that hinder the movement of foodstuffs among these nations.

In Japan, the Food Chemistry Division of the Ministry of Health and Welfare (MWH) has jurisdiction over food additives through the Food Sanitation Law (www.mhlw.go.jp/english/index.htm), which was enacted in January 1948 with several amendments adopted since then. Amendments to the regulations as well as additives or deletions to Kohetisho (Japanese Codex of Food Additives) were mostly influenced by two major objectives.

1. Protection of food sanitation and customer safety.

2. Harmonization with international regulatory requirements.

Under the amended Japanese Food Sanitation Law (1995), substances that are permitted for food use as food additives fall into four categories:

1. Substances that are generally recognized as safe (~ 700 substances).

2. Natural-based flavours (~ 580 substances).

3. Natural-based substances that are recognized as safe to human health on the basis of actual results for use as food additives (~ 490 substances).

4. Synthetic substances that are recognized as safe to human health on the basis of the results of safety evaluations (~ 350 substances).

The amended law requires that any new substances, regardless of whether they are natural-based or synthetic, be verified to be safe to human health through safety evaluation and then approved by MHW before being used as food additives. Also, it is required that the type of substance and the intended purpose of addition be labeled on the surface of food containers or packaging.

FOOD ADDITIVES THAT DO NOT REQUIRE CERTIFICATION

- Annatto extract
- Beet powder
- Canthaxanthin
- Carmine
- Cochineal extract
- Ferrous gluconate
- Grape skins extract
- Riboflavin
- Titanium dioxide
- Ultramarine blue

- β-Carotene
- β-Apo-8-carotenal
- Caramel
- Carrot oil
- Cottonseed flour
- Fruit, Vegetable juices
- Paprika, Paprika oleoresin
- Saffron
- Turmeric, turmeric oleoresin

TASTE SUBSTANCES

- Sweet
- Bitter
- Sour
- Salty

Chemically cations cause salty tastes and anions inhibits salty tastes.

UMAMI TASTE

Much of the flavour contributed to food by yeast hydrolysates results from the 5'-ribonucleotides present. It is well documented that a synergistic interaction occurs between monosodium L-glutamate (MSG) and the 5'-ribonucleotides in providing both the umami taste and in enhancing flavours.

CHELATING AGENTS (SEQUESTRANTS)

It play a significant role in food stabilization through reactions with metallic and alkaline earth ions to form complexes that alter the properties of the ions and their effects in foods. Many metals exist in a naturally chelated state. Examples include magnesium in chlorophyll, copper, iron, zinc and manganese in various enzymes, iron in proteins such as ferritin and iron in the porphyrin ring of myoglobin and hemoglobin. Citric acid and its derivatives, various phosphates and salts of ethylene diamine tetra acetic acid (EDTA) are the most popular chelating agents used in food.

ANTIMICROBIAL AGENTS

- Sulfites and sulfur dioxide

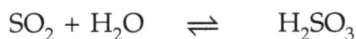

$$SO_2 + H_2O \rightleftharpoons H_2SO_3$$

$$2H_2SO_3 \rightleftharpoons H^+, HSO^-_3 + 2H^+, SO^{2-}_3$$

- Nitrite and nitrate salts
- Sorbic acid (C – C = C – C = C – COOH)
- Natamycin
- Glyeryl esters
- Propionic acid ($CH_3 – CH_2 – COOH$)
- Acetic acid ($CH_3 – COOH$)
- Benzoic acid (C_6H_5COOH)

- p-Hydeoxybenzoate alkyl esters
- Epoxides
- Antibiotics
- Diethyl pyrocarbonate

NON-NUTRITIVE AND LOW-CALORIE SWEETENERS

- Cyclamates
- Saccharin
- Aspartame
- Acesulfamek
- Others
  - Glycyrrhizic
  - Stevia rebaudiana Bertoni
  - Neohesperidin dihydrochalcone
  - Thaumatococcus daniellii
  - Miraculin

FLOUR BLEACHING AGENTS

- Benzoyl peroxide $[(C_6H_5CO)_2O_2]$
- Chlorine gas (Cl_2)
- Chlorine dioxide (ClO_2)
- Nitrosyl chloride $(NOCl)$
- Nitrogen dioxide (NO_2)
- Nitrogen tetroxide (N_2O_4)

DOUGH IMPROVERS AND CONDITIONERS

- Potassium bromate $(KBrO_3)$
- Potassium iodate (KIO_3)
- Calcium iodate $[Ca(IO_3)2]$
- Calcium peroxide (CaO_2)
- Ammonium chloride (NH_4Cl)
- Ammonium sulphate $[(NH_4)_2SO_4]$

- Calcium sulphate ($CaSO_4$)
- Ammonium phosphate [$(NH_4)_3PO_4$]
- Calcium phosphate ($CaHPO_4$)
- Calcium stearoyl-2-lactylate

ANTICAKING AGENTS

- Calcium silicate
- Calcium and magnesium salts of long-chain fatty acids.
- Calcium stearate
- Sodium silicoaluminate
- Tricalcium phosphate
- Magnesium silicate
- Magnesium carbonate
- Microcrystalline cellulose powder

PROPELLANTS USED IN FOOD INDUSTRY

- Nitrous oxide
- Nitrogen gas
- Carbon dioxide
- Octafluorocyclobutane (Freon C-318)
- Chloropentafluoroethane (Freon C-115)

E NUMBER

- 1 Classification by numeric range
- 2 Full list
  - 2.1 E100-E199 (colours)
  - 2.2 E200-E299 (preservatives)
  - 2.3 E300-E399 (antioxidants, acidity regulators)
  - 2.4 E400-E499 (thickeners, stabilizers, emulsifiers)
  - 2.5 E500-E599 (acidity regulators, anti-caking agents)
  - 2.6 E600-E699 (flavour enhancers)
  - 2.7 E700-E799 (antibiotics)

- 2.8 E900-E999 (miscellaneous)
- 2.9 E1000-E1999 (additional chemicals)

1. Classification by numeric range

100-199 Colours	100-109	yellows (see the E number Full list)
	110-119	oranges
	120-129	reds
	130-139	blues & violets
	140-149	greens
	150-159	browns & blacks
	160-199	others
200-299 Preservatives	200-209	sorbates
	210-219	benzoates
	220-229	sulphites
	230-239	phenols & formates (methanoates)
	240-259	nitrates
	260-269	acetates (ethanoates)
	270-279	lactates
	280-289	propionates (propanoates)
	290-299	others
300-399 Antioxidants & acidity regulators	300-305	ascorbates (vitamin C)
	306-309	Tocopherol (vitamin E)
	310-319	gallates & erythorbates
	320-329	lactates
	330-339	citrates & tartrates
	340-349	phosphates
	350-359	malates & adipates
	360-369	succinates & fumarates
	370-399	others
400-499 Thickeners, stabilizers & emulsifiers	400-409	alginates
	410-419	natural gums
	420-429	other natural agents
	430-439	polyoxyethene compounds
	440-449	natural emulsifiers
	450-459	phosphates
	460-469	cellulose compounds
	470-489	fatty acids & compounds
	490-499	others

500-599 pH regulators & anti-caking agents	500-509	mineral acids & bases	
	510-519	chlorides & sulphates	
	520-529	sulphates & hydroxides	
	530-549	alkali metal compounds	
	550-559	silicates	
	570-579	stearates & gluconates	
	580-599	others	
600-699 Flavour enhancers	620-629	glutamates	
	630-639	inosinates	
	640-649	others	
700-799 Antibiotics	710-713		
900-999 Miscellaneous	900-909	waxes	
	910-919	synthetic glazes	
	920-929	improving agents	
	930-949	packaging gases	
	950-969	sweeteners	
	990-999	foaming agents	
1100-1599 Additional chemicals	New chemicals that do not fall into standard classification schemes		

NB : Not all examples of a class fall into the given numeric range. Moreover, many chemicals, particularly in the E400-499 range, have a variety of purposes.

FULL LIST

Each additional has its status:

- permitted additionals are labeled with N/A;
- forbidden additionals were proven to cause diseases beyond any doubt;
- unpermitted additionals are those for which conclusive test data is not yet available either due to ongoing tests or no testing;
- dangerous additionals may be dangerous for people with chronic diseases.

E100-E199 (colours)

Code	Name	Purpose
E100	Curcumin, turmeric	food colouring (yellow-orange)
E101	Riboflavin (Vitamin B_2), formerly called lactoflavin (Vitamin G)	food colouring (yellow-orange)
E101a	Riboflavin-5'-Phosphate	food colouring (yellow-orange)
E102	Tartrazine (FD&C Yellow 5)	food colouring (lemon yellow)

Contd...

Code	Name	Purpose
E103	Chrysoine resorcinol	food colouring (golden)
E104	Quinoline Yellow WS	food colouring (dull or greenish yellow)
E105	Fast Yellow AB	food colouring (yellow)
E106	Riboflavin-5-Sodium Phosphate	food colouring (yellow)
E107	Yellow 2G	food colouring (yellow)
E110	Sunset Yellow FCF (Orange Yellow S, FD&C Yellow 6)	food colouring (yellow-orange)
E111	Orange GGN	food colouring (orange)
E120	Cochineal, Carminic acid, Carmines, Natural Red 4	food colouring (crimson)
E121	Citrus Red 2	food colouring (dark red)
E122	Carmoisine, Azorubine	food colouring (red to maroon)
E123	Amaranth (FD&C Red 2)	food colouring (dark red)
E124	Ponceau 4R (Cochineal Red A, Brilliant Scarlet 4R)	food colouring (red)
E125	Ponceau SX, Scarlet GN	food colouring (red)
E126	Ponceau 6R	food colouring (red)
E127	Erythrosine (FD&C Red 3)	food colouring (red)
E128	Red 2G	food colouring (red)
E129	Allura Red AC (FD&C Red 40)	food colouring (red)
E130	Indanthrene blue RS	food colouring (blue)
E131	Patent Blue V	food colouring (dark blue)
E132	Indigo carmine, FD&C Blue 2	food colouring (indigo)
E133	Brilliant Blue FCF (FD&C Blue 1)	food colouring (reddish blue)
E140	Chlorophylls and Chlorophyllins: (i) Chlorophylls (ii) Chlorophyllins	food colouring (green)
E141	Copper complexes of chlorophylls and chlorophyllins (i) Copper complexes of chlorophylls (ii) Copper complexes of chlorophyllins	food colouring (green)
E142	Green S	food colouring (green)
E143	Fast Green FCF (FD&C Green 3)	food colouring (sea green)
E150a	Plain caramel	food colouring
E150b	Caustic sulfite caramel	food colouring
E150c	Ammonia caramel	food colouring
E150d	Sulphite ammonia caramel	food colouring

Contd...

Code	Name	Purpose
E151	Black PN, Brilliant Black BN	food colouring
E152	Black 7984	food colouring
E153	Carbon black, Vegetable carbon	food colouring
E154	Brown FK, Kipper Brown	food colouring
E155	Brown HT, Chocolate brown HT	food colouring
E160a	Alpha-carotene, Beta-carotene, Gamma-carotene	food colouring
E160b	Annatto, bixin, norbixin	food colouring
E160c	Paprika extract, Capsanthin, capsorubin	food colouring
E160d	Lycopene	food colouring
E160e	Beta-apo-8'-carotenal (C 30)	food colouring
E160f	Ethyl ester of beta-apo-8'-carotenic acid (C 30)	food colouring
E161a	Flavoxanthin	food colouring
E161b	Lutein	food colouring
E161c	Cryptoxanthin	food colouring
E161d	Rubixanthin	food colouring
E161e	Violaxanthin	food colouring
E161f	Rhodoxanthin	food colouring
E161g	Canthaxanthin	food colouring
E161h	Zeaxanthin	food colouring
E161i	Citranaxanthin	food colouring
E161j	Astaxanthin	food colouring
E162	Beetroot Red, Betanin	food colouring
E163	Anthocyanins	food colouring
E170	Calcium carbonate, Chalk	food colouring
E171	Titanium dioxide	food colouring (pure white)
E172	Iron oxides and hydroxides	food colouring
E173	Aluminium	food colouring
E174	Silver	food colouring
E175	Gold	food colouring
E180	Pigment Rubine, Lithol Rubine BK	food colouring
E181	Tannin	food colouring
E182	Orcein, Orchil	food colouring

E200–E299 (preservatives)

Code	Name	Purpose
E200	Sorbic acid	preservative
E201	Sodium sorbate	preservative
E202	Potassium sorbate	preservative
E203	Calcium sorbate	preservative
E209	Heptyl p-hydroxybenzoate	preservative
E210	Benzoic acid	preservative
E211	Sodium benzoate	preservative
E212	Potassium benzoate	preservative
E213	Calcium benzoate	preservative
E214	Ethylparaben (ethyl para-hydroxybenzoate)	preservative
E215	Sodium ethyl para-hydroxybenzoate	preservative
E216	Propylparaben (propyl para-hydroxy benzoate)	preservative
E217	Sodium propyl para-hydroxybenzoate	preservative
E218	Methylparaben (methyl para-hydroxy benzoate)	preservative
E219	Sodium methyl para-hydroxybenzoate	preservative
E220	Sulphur dioxide	preservative
E221	Sodium sulphite	preservative
E222	Sodium bisulphite (sodium hydrogen sulphite)	preservative
E223	Sodium metabisulphite	preservative
E224	Potassium metabisulphite	preservative
E225	Potassium sulphite	preservative
E226	Calcium sulphite	preservative
E227	Calcium hydrogen sulphite (preservative)	firming agent
E228	Potassium hydrogen sulphite	preservative
E230	Biphenyl, diphenyl	preservative
E231	Orthophenyl phenol	preservative
E232	Sodium orthophenyl phenol	preservative
E233	Thiabendazole	preservative
E234	Nisin	preservative
E235	Natamycin, Pimaracin	preservative
E236	Formic acid	preservative

Contd...

Code	Name	Purpose
E237	Sodium formate	preservative
E238	Calcium formate	preservative
E239	Hexamine (hexamethylene tetramine)	preservative
E240	Formaldehyde	preservative
E242	Dimethyl dicarbonate	preservative
E249	Potassium nitrite	preservative
E250	Sodium nitrite	preservative
E251	Sodium nitrate	preservative
E252	Potassium nitrate (Saltpetre)	preservative
E260	Acetic acid (preservative)	acidity regulator
E261	Potassium acetate (preservative)	acidity regulator
E262	Sodium acetates (i) Sodium acetate (ii) Sodium hydrogen acetate (sodium diacetate)	preservative, acidity regulator
E263	Calcium acetate (preservative)	acidity regulator
E264	Ammonium acetate	preservative
E265	Dehydroacetic acid	preservative
E266	Sodium dehydroacetate	preservative
E270	Lactic acid (preservative) (acid)	antioxidant
E280	Propionic acid	preservative
E281	Sodium propionate	preservative
E282	Calcium propionate	preservative
E283	Potassium propionate	preservative
E284	Boric acid	preservative
E285	Sodium tetraborate (borax)	preservative
E290	Carbon dioxide	acidity regulator
E296	Malic acid (acid)	acidity regulator
E297	Fumaric acid	acidity regulator

E300-E399 (antioxidants, acidity regulators)

Code	Name	Purpose
E300	Ascorbic acid (Vitamin C)	antioxidant
E301	Sodium ascorbate	antioxidant
E302	Calcium ascorbate	antioxidant
E303	Potassium ascorbate	antioxidant

Contd...

Code	Name	Purpose
E304	Fatty acid esters of ascorbic acid (i) Ascorbyl palmitate	antioxidant
E305	(ii) Ascorbyl stearate	antioxidant
E306	Tocopherol-rich extract (natural)	antioxidant
E307	Alpha-tocopherol (synthetic)	antioxidant
E308	Gamma-tocopherol (synthetic)	antioxidant
E309	Delta-tocopherol (synthetic)	antioxidant
E310	Propyl gallate	antioxidant
E311	Octyl gallate	antioxidant
E312	Dodecyl gallate	antioxidant
E313	Ethyl gallate	antioxidant
E314	Guaiac resin	antioxidant
E315	Erythorbic acid	antioxidant
E316	Sodium erythorbate	antioxidant
E317	Erythorbin acid	antioxidant
E318	Sodium erythorbin	antioxidant
E319	tert-Butylhydroquinone (TBHQ)	antioxidant
E320	Butylated hydroxyanisole (BHA)	antioxidant
E321	Butylated hydroxytoluene (BHT)	antioxidant
E322	Lecithin	emulsifier
E323	Anoxomer	antioxidant
E324	Ethoxyquin	antioxidant
E325	Sodium lactate	acidity regulator
E326	Potassium lactate (antioxidant)	acidity regulator
E327	Calcium lactate	acidity regulator
E328	Ammonium lactate	acidity regulator
E329	Magnesium lactate	acidity regulator
E330	Citric acid	acid, acidity regulator
E331	Sodium citrates (i) Monosodium citrate (ii) Disodium citrate (iii) Sodium citrate (trisodium citrate)	acidity regulator
E332	Potassium citrates (i) Monopotassium citrate (ii) Potassium citrate (tripotassium citrate)	acidity regulator
E333	Calcium citrates (i) Monocalcium citrate (ii) Dicalcium citrate (iii) Calcium citrate (tricalcium citrate)	acidity regulator, firming agent, sequestrant

Contd...

Code	Name	Purpose
E334	Tartaric acid (L(+)-)	acid
E335	Sodium tartrates (i) Monosodium tartrate (ii), Disodium tartrate	acidity regulator
E336	Potassium tartrates (i) Monopotassium tartrate (cream of tartar) (ii) Dipotassium	tartrate acidity regulator
E337	Sodium potassium tartrate	acidity regulator
E338	Orthophosphoric acid	acid
E339	Sodium phosphates (i) Monosodium phosphate (ii) Disodium phosphate (iii) Trisodium phosphate	antioxidant
E340	Potassium phosphates (i) Monopotassium phosphate (ii) Dipotassium phosphate (iii) Tripotassium phosphate	antioxidant
E341	Calcium phosphates (i) Monocalcium phosphate (ii) Dicalcium phosphate (iii) Tricalcium phosphate	anti-caking agent, firming agent
E342	Ammonium phosphates: (i) Monoammonium orthophosphate (ii) Diammonium orthophosphate	
E343	Magnesium phosphates (i) monomagnesium phosphate (ii) Dimagnesium phosphate	anti-caking agent
E344	Lecitin citrate	acidity regulator
E345	Magnesium citrate	acidity regulator
E349	Ammonium malate	acidity regulator
E350	Sodium malates (i) Sodium malate (ii) Sodium hydrogen malate	acidity regulator
E351	Potassium malate	acidity regulator
E352	Calcium malates (i) Calcium malate (ii) Calcium hydrogen malate	acidity regulator
E353	Metatartaric acid	emulsifier
E354	Calcium tartrate	emulsifier
E355	Adipic acid	acidity regulator

Contd...

Code	Name	Purpose
E356	Sodium adipate	acidity regulator
E357	Potassium adipate	acidity regulator
E359	Ammonium adipate	acidity regulator
E363	Succinic acid	acidity regulator
E365	Sodium fumarate	acidity regulator
E366	Potassium fumarate	acidity regulator
E367	Calcium fumarate	acidity regulator
E368	Ammonium fumarate	acidity regulator
E370	1,4-Heptonolactone	acidity regulator
E375	Niacin (nicotinic acid), Nicotinamide	colour retention agent
E380	Triammonium citrate	acidity regulator
E381	Ammonium ferric citrate	acidity regulator
E383	Calcium glycerylphosphate	acidity regulator
E384	Isopropyl citrate	acidity regulator
E385	Calcium disodium ethylene diamine tetraacetate, (Calcium disodium EDTA)	sequestrant
E386	Disodium ethylene diamine tetraacetate (Disodium EDTA)	sequestrant
E387	Oxystearin	stabiliser
E388	Thiodipropionic acid	
E389	Dilauryl thiodipropionate	
E390	Distearyl thiodipropionate	
E391	Phytic acid	
E399	Calcium lactobionate	

E400-E499 (thickeners, stabilizers, emulsifiers)

Code	Name	Purpose
E400	Alginic acid (thickener) (stabiliser) (gelling agent)	emulsifier
E401	Sodium alginate (thickener) (stabiliser) (gelling agent)	emulsifier
E402	Potassium alginate (thickener) (stabiliser) (gelling agent)	emulsifier
E403	Ammonium alginate (thickener) (stabiliser)	emulsifier
E404	Calcium alginate (thickener) (stabiliser) (gelling agent)	emulsifier

Contd...

Code	Name	Purpose
E405	Propane-1,2-diol alginate (Propylene glycol alginate) (thickener) (stabiliser)	emulsifier
E406	Agar (thickener) (gelling agent)	stabiliser
E407	Carrageenan (thickener) (stabiliser) (gelling agent)	emulsifier
E407a	Processed eucheuma seaweed (thickener) (stabiliser) (gelling agent)	emulsifier
E408	Bakers yeast glycan	
E409	Arabinogalactan	
E410	Locust bean gum (Carob gum) (thickener) (stabiliser) (gelling agent)	emulsifier
E411	Oat gum (thickener)	stabiliser
E412	Guar gum (thickener)	stabiliser
E413	Tragacanth (thickener) (stabiliser)	emulsifier
E414	Acacia gum (gum arabic) (thickener) (stabiliser)	emulsifier
E415	Xanthan gum (thickener)	stabiliser
E416	Karaya gum (thickener) (stabiliser)	emulsifier
E417	Tara gum (thickener)	stabiliser
E418	Gellan gum (thickener) (stabiliser)	emulsifier
E419	Gum ghatti (thickener) (stabiliser)	emulsifier
E420	Sorbitol (i) Sorbitol (ii) Sorbitol syrup (emulsifier) (sweetener)	humectant
E421	Mannitol (anti-caking agent)	sweetener
E422	Glycerol (emulsifier)	sweetener
E424	Curdlan	
E425	Konjac (i) Konjac gum (ii) Konjac glucomannane	emulsifier
E426	Soybean hemicellulose	
E429	Peptones	
E430	Polyoxyethene (8) stearate (emulsifier)	stabiliser
E431	Polyoxyethene (40) stearate	emulsifier
E432	Polyoxyethene (20) sorbitan monolaurate (polysorbate 20)	emulsifier
E433	Polyoxyethene (20) sorbitan monooleate (polysorbate 80)	emulsifier
E434	Polyoxyethene (20) sorbitan monopalmitate (polysorbate 40)	emulsifier

Contd...

Code	Name	Purpose
E435	Polyoxyethene (20) sorbitan monostearate (polysorbate 60)	emulsifier
E436	Polyoxyethene (20) sorbitan tristearate (polysorbate 65)	emulsifier
E440	Pectins (i) pectin (ii) amidated pectin	emulsifier
E441	Gelatine (emulsifier)	gelling agent
E442	Ammonium phosphatides	emulsifier
E443	Brominated vegetable oil	
E444	Sucrose acetate isobutyrate	emulsifier
E445	Glycerol esters of wood rosins	emulsifier
E446	Succistearin	
E450	Diphosphates (i) Disodium diphosphate (ii) Trisodium diphosphate (iii) Tetrasodium diphosphate (iv) Dipotassium diphosphate (v) Tetrapotassium diphosphate (vi) Dicalcium diphosphate (vii) Calcium dihydrogen diphosphate	emulsifier
E451	Triphosphates (i) Sodium tripolyphosphate (pentasodium triphosphate) (ii) Pentapotassium triphosphate	emulsifier
E452	Polyphosphates (i) Sodium polyphosphates (ii) Potassium polyphosphates (iii) Sodium calcium polyphosphate (iv) Calcium polyphosphates (v) Ammonium Polyphosphate	emulsifier
E459	Beta-cyclodextrine	emulsifier
E460	Cellulose (i) Microcrystalline cellulose (ii) Powdered cellulose	emulsifier
E461	Methyl cellulose	emulsifier
E462	Ethyl cellulose	emulsifier
E463	Hydroxypropyl cellulose	emulsifier
E464	Hydroxy propyl methyl cellulose, = hypromellose	emulsifier
E465	Ethyl methyl cellulose	emulsifier
E466	Carboxymethyl cellulose, Sodium carboxy methyl cellulose	emulsifier

Contd...

Code	Name	Purpose
E467	Ethyl hydroxyethyl cellulose	
E468	Crosslinked sodium carboxymethyl cellulose (Croscarmellose)	emulsifier
E469	Enzymically hydrolysed carboxymethyl cellulose	emulsifier
E470	Salts of fatty acids with base Al, Ca, Na, Mg, K and NH_4	
E470a	Sodium, potassium and calcium salts of fatty acids (emulsifier)	anti-caking agent
E470b	Magnesium salts of fatty acids (emulsifier)	anti-caking agent
E471	Mono- and diglycerides of fatty acids (glyceryl monostearate, glyceryl distearate)	emulsifier
E472a	Acetic acid esters of mono- and diglycerides of fatty acids	emulsifier
E472b	Lactic acid esters of mono- and diglycerides of fatty acids	emulsifier
E472c	Citric acid esters of mono- and diglycerides of fatty acids	emulsifier
E472d	Tartaric acid esters of mono- and diglycerides of fatty acids	emulsifier
E472e	Mono- and diacetyl tartaric acid esters of mono- and diglycerides of fatty acids	emulsifier
E472f	Mixed acetic and tartaric acid esters of mono- and diglycerides of fatty acids	emulsifier
E472g	Succinylated monoglycerides	emulsifier
E473	Sucrose esters of fatty acids	emulsifier
E474	Sucroglycerides	emulsifier
E475	Polyglycerol esters of fatty acids	emulsifier
E476	Polyglycerol polyricinoleate	emulsifier
E477	Propane-1, 2-diol esters of fatty acids, propylene glycol esters of fatty acids	emulsifier
E478	Lactylated fatty acid esters of glycerol and propane-1	emulsifier
E479b	Thermally oxidized soya bean oil interacted with mono- and diglycerides of fatty acids	emulsifier
E480	Dioctyl sodium sulphosuccinate	emulsifier
E481	Sodium stearoyl-2-lactylate	emulsifier
E482	Calcium stearoyl-2-lactylate	emulsifier

Contd...

Code	Name	Purpose
E483	Stearyl tartrate	emulsifier
E484	Stearyl citrate	emulsifier
E485	Sodium Stearoyl Fumarate	emulsifier
E486	Calcium Stearoyl Fumarate	emulsifier
E487	Sodium laurylsulphate	emulsifier
E488	Ethoxylated Mono- and Di-Glycerides	emulsifier
E489	Methyl Glucoside - Coconut Oil Ester	emulsifier
E490	Propane-1,2-diol	
E491	Sorbitan monostearate	emulsifier
E492	Sorbitan tristearate	emulsifier
E493	Sorbitan monolaurate	emulsifier
E494	Sorbitan monooleate	emulsifier
E495	Sorbitan monopalmitate	emulsifier
E496	Sorbitan trioleat	emulsifier
E497	Polyoxypropylene-polyoxyethylene polymers	
E498	Partial polyglycerol esters of polycondensed fatty acids of castor oil	
E499	Cassia gum	

E500-E599 (acidity regulators, anti-caking agents)

Code	Name	Purpose
E500	Sodium carbonates (i) Sodium carbonate (ii) Sodium bicarbonate (Sodium hydrogen carbonate) (iii) Sodium sesquicarbonate (acidity regulator)	raising agent
E501	Potassium carbonates (i) Potassium carbonate (ii) Potassium bicarbonate (Potassium hydrogen carbonate)	acidity regulator
E503	Ammonium carbonates (i) Ammonium carbonate (ii) Ammonium bicarbonate (Ammonium hydrogen carbonate)	acidity regulator
E504	Magnesium carbonates (i) Magnesium carbonate (ii) Magnesium bicarbonate Magnesium hydrogen carbonate	acidity regulator, anti-caking agent

Contd...

Code	Name	Purpose
E505	Ferrous carbonate	acidity regulator
E507	Hydrochloric acid	acid
E508	Potassium chloride (gelling agent)	seasoning
E509	Calcium chloride (sequestrant)	firming agent
E510	Ammonium chloride, ammonia solution (acidity regulator)	improving agent
E511	Magnesium chloride	firming agent
E512	Stannous chloride	antioxidant
E513	Sulphuric acid	acid
E514	Sodium sulphates (i) Sodium sulphate (ii)	acid
E515	Potassium Sulphates (i) Potassium Sulphate (ii)	
E516	Calcium sulphate	
E517	Ammonium sulphate	improving agent
E518	Magnesium sulfate (Epsom salts), (acidity regulator)	firming agent
E519	Copper (II) sulphate	preservative
E520	Aluminium sulphate	firming agent
E521	Aluminium sodium sulphate	firming agent
E522	Aluminium potassium sulphate	acidity regulator
E523	Aluminium ammonium sulphate	acidity regulator
E524	Sodium hydroxide	acidity regulator
E525	Potassium hydroxide	acidity regulator
E526	Calcium hydroxide (acidity regulator)	firming agent
E527	Ammonium hydroxide	acidity regulator
E528	Magnesium hydroxide	acidity regulator
E529	Calcium oxide (acidity regulator)	improving agent
E530	Magnesium oxide (acidity regulator)	anti-caking agent
E535	Sodium ferrocyanide (acidity regulator)	anti-caking agent
E536	Potassium ferrocyanide	anti-caking agent
E537	Ferrous hexacyanomanganate	anti-caking agent
E538	Calcium ferrocyanide	anti-caking agent
E539	Sodium thiosulphate	antioxidant
E540	Dicalcium diphosphate (acidity regulator)	emulsifier
E541	Sodium aluminium phosphate (i) Acidic (ii) Basic	emulsifier
E542	Bone phosphate (Essentiale Calcium Phosphate, Tribasic)	anti-caking agent

Contd...

Code	Name	Purpose
E543	Calcium sodium polyphosphate	emulsifier
E544	Calcium polyphosphate	emulsifier
E545	Ammonium polyphosphate	emulsifier
E550	Sodium Silicates (i) Sodium silicate (ii) Sodium metasilicate	anti-caking agent
E551	Silicon dioxide (Silica)	anti-caking agent
E552	Calcium silicate	anti-caking agent
E553a	(i) Magnesium silicate (ii) Magnesium trisilicate	anti-caking agent
E553b	Talc	anti-caking agent
E554	Sodium aluminosilicate (sodium aluminium silicate)	anti-caking agent
E555	Potassium aluminium silicate	anti-caking agent
E556	Calcium aluminosilicate (calcium aluminium silicate)	anti-caking agent
E557	Zinc silicate	anti-caking agent
E558	Bentonite	anti-caking agent
E559	Aluminium silicate (Kaolin)	anti-caking agent
E560	Potassium silicate	anti-caking agent
E561	Vermiculite	
E562	Sepiolite	
E563	Sepiolitic clay	
E565	Lignosulphonates	
E566	Natrolite-phonolite	
E570	Stearic acid (Fatty acid)	anti-caking agent
E572	Magnesium stearate, calcium stearate (emulsifier)	anti-caking agent
E574	Gluconic acid	acidity regulator
E575	Glucono delta-lactone (acidity regulator)	sequestrant
E576	Sodium gluconate	sequestrant
E577	Potassium gluconate	sequestrant
E578	Calcium gluconate	firming agent
E579	Ferrous gluconate	food colouring
E580	Magnesium gluconate	
E585	Ferrous lactate	food colouring
E586	4-Hexylresorcinol	antioxidant
E598	Synthetic calcium aluminates	
E599	Perlite	

E600-E699 (flavour enhancers)

Code	Name	Purpose
E620	Glutamic acid	flavour enhancer
E621	Monosodium glutamate (MSG)	flavour enhancer
E622	Monopotassium glutamate	flavour enhancer
E623	Calcium diglutamate	flavour enhancer
E624	Monoammonium glutamate	flavour enhancer
E625	Magnesium diglutamate	flavour enhancer
E626	Guanylic acid	flavour enhancer
E627	Disodium guanylate, sodium guanylate	flavour enhancer
E628	Dipotassium guanylate	flavour enhancer
E629	Calcium guanylate	flavour enhancer
E630	Inosinic acid	flavour enhancer
E631	Disodium inosinate	flavour enhancer
E632	Dipotassium inosinate	flavour enhancer
E633	Calcium inosinate	flavour enhancer
E634	Calcium 5'-ribonucleotides	flavour enhancer
E635	Disodium 5'-ribonucleotides	flavour enhancer
E636	Maltol	flavour enhancer
E637	Ethyl maltol	flavour enhancer
E640	Glycine and its sodium salt	flavour enhancer
E641	L-leucine	flavour enhancer
E642	Lysine hydrochloride	flavour enhancer
E650	Zinc acetate	flavour enhancer

E700-E799 (antibiotics)

Code	Name	Purpose
E700	Bacitracin	
E701	Tetracyclines	
E710	Spiramycins	
E711	Virginiamicins	
E712	Flavophospholipol	
E713	Tylosin	
E714	Monensin	
E715	Avoparcin	
E716	Salinomycin	
E717	Avilamycin	

E900-E999 (miscellaneous)

Code	Name	Purpose
E900	Dimethyl polysiloxane (anti-foaming agent)	anti-caking agent
E901	Beeswax , white and yellow	glazing agent
E902	Candelilla wax	glazing agent
E903	Carnauba wax	glazing agent
E904	Shellac	glazing agent
E905	Paraffins	
E905a	Mineral oil	anti-foaming agent
E905b	Petrolatum	
E905c	Petroleum wax (i) Microcrystalline wax (ii) Paraffin wax	glazing agent
E906	Gum benzoic	flavour enhancer
E907	Crystalline wax	glazing agent
E908	Rice bran wax	glazing agent
E909	Spermaceti wax	glazing agent
E910	Wax esters	glazing agent
E911	Methyl esters of fatty acids	glazing agent
E912	Montan wax	glazing agent
E913	Lanolin, sheep wool grease	glazing agent
E914	Oxidized polyethylene wax	glazing agent
E915	Esters of colophony	glazing agent
E916	Calcium iodate	
E917	Potassium iodate	
E918	Nitrogen oxides	
E919	Nitrosyl chloride	
E920	L-cysteine	improving agent
E921	L-cystine	improving agent
E922	Potassium persulfate	improving agent
E923	Ammonium persulfate	improving agent
E924	Potassium bromate	improving agent
E924b	Calcium bromate	improving agent
E925	Chlorine	preservative, bleach, improving agent
E926	Chlorine dioxide (preservative)	bleach
E927a	Azodicarbonamide	improving agent
E927b	Carbamide (urea)	improving agent
E928	Benzoyl peroxide (improving agent)	bleach

Contd...

Code	Name	Purpose
E929	Acetone peroxide	
E930	Calcium peroxide (improving agent)	bleach
E938	Argon	packaging gas
E939	Helium	packaging gas
E940	Dichlorodifluoromethane	packaging gas
E941	Nitrogen (packaging gas)	propellant
E942	Nitrous oxide	propellant
E943a	Butane	propellant
E943b	Isobutane	propellant
E944	Propane	propellant
E945	Chloropentafluoroethane	propellant
E946	Octafluorocyclobutane	propellant
E948	Oxygen	packaging gas
E949	Hydrogen	packaging gas
E950	Acesulfame potassium	sweetener
E951	Aspartame	sweetener
E952	Cyclamic acid and its sodium and calcium salts, also known as Cyclamate	sweetener
E953	Isomalt, Isomaltitol	sweetener
E954	Saccharin and its sodium, potassium and calcium salts	sweetener
E955	Sucralose (Trichlorogalactosucrose)	sweetener
E956	Alitame	sweetener
E957	Thaumatin (sweetener)	flavour enhancer
E958	Glycyrrhizin (sweetener)	flavour enhancer
E959	Neohesperidine dihydrochalcone (sweetener)	flavour enhancer
E960	Stevioside	sweetener
E961	Neotame	sweetener
E962	Aspartame-acesulfame salt (sweetener)	stabiliser
E965	Maltitol (i) Maltitol (ii) Maltitol syrup (sweetener) (stabiliser)	humectant
E966	Lactitol	sweetener
E967	Xylitol	sweetener
E968	Erythritol	humectant
E999	Quillaia extract	foaming agent

E1000–E1999 (additional chemicals)

Code	Name	Purpose
E1000	Cholic acid	emulsifier
E1001	Choline salts	emulsifier
E1100	Amylase	stabiliser, flavour enhancer
E1101	Proteases ((i) Protease, (ii) Papain, (iii) Bromelain, (iv) Ficin)	stabiliser, flavour enhancer
E1102	Glucose oxidase	antioxidant
E1103	Invertase	stabiliser
E1104	Lipases	
E1105	Lysozyme	preservative
E1200	Polydextrose	stabiliser, thickening agent, humectant, carrier
E1201	Polyvinylpyrrolidone	stabiliser
E1202	Polyvinylpolypyrrolidone (carrier)	stabiliser
E1203	Polyvinyl alcohol	
E1204	Pullulan	
E1400	Dextrin (Dextrins, roasted starch white and yellow) (stabiliser)	thickening agent
E1401	Modified starch ((Acid-treated starch) stabiliser)	thickening agent
E1402	Alkaline modified starch (stabiliser)	thickening agent
E1403	Bleached starch (stabiliser)	thickening agent
E1404	Oxidized starch (emulsifier)	thickening agent
E1405	Enzyme treated starch	
E1410	Monostarch phosphate (stabiliser)	thickening agent
E1411	Distarch glycerol (thickening agent)	emulsifier
E1412	Distarch phosphate esterified with sodium trimetasphosphate; esterified with phosphorus oxychloride (stabiliser)	thickening agent
E1413	Phosphated distarch phosphate (stabiliser)	thickening agent
E1414	Acetylated distarch phosphate (emulsifier)	thickening agent
E1420	Starch acetate esterified with acetic anhydride (stabiliser)	thickening agent
E1421	Starch acetate esterified with vinyl acetate (stabiliser)	thickening agent
E1422	Acetylated Distarch Adipate (stabiliser)	thickening agent
E1423	Acetylated distarch glycerol	thickening agent
E1430	Distarch glycerine (stabiliser)	thickening agent
E1440	Hydroxy propyl starch (emulsifier)	thickening agent

Contd...

Code	Name	Purpose
E1441	Hydroxy propyl distarch glycerine (stabiliser)	thickening agent
E1442	Hydroxy propyl distarch phosphate (stabiliser)	thickening agent
E1443	Hydroxy propyl distarch glycerol	
E1450	Starch sodium octenyl succinate (emulsifier) (stabiliser)	thickening agent
E1451	Acetylated oxidised starch (emulsifier)	thickening agent
E1501	Benzylated hydrocarbons	
E1502	Butane-1, 3-diol	
E1503	Castor oil	resolving agent
E1504	Ethyl acetate	flavour solvent
E1505	Triethyl citrate	foam stabiliser
E1510	Ethanol	
E1516	Glyceryl monoacetate	flavour solvent
E1517	Glyceryl diacetate or diacetin	flavour solvent
E1518	Glyceryl triacetate or triacetin	humectant
E1519	Benzyl alcohol	
E1520	Propylene glycol	humectant
E1521	Polyethylene glycerol	
E1525	Hydroxyethyl cellulose	thickening agent

HYSTERESIS

A moisture sorption isotherm prepared by the addition of water (resorption) to a dry sample will not necessarily be super imposable with an isotherm prepared by desorption. This lack of superimposibility is referred to as hysteresis and schematic example is shown in following figure.

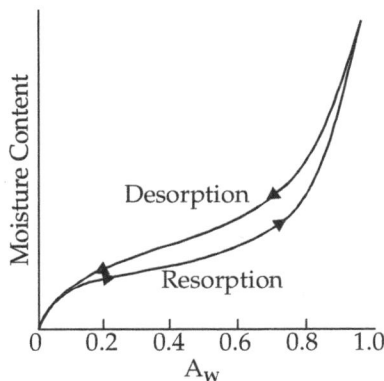

PROTEIN CLASSIFICATION

- Simple Proteins
  - Globular proteins
  - Albumins
  - Globulins
  - Glutelins
  - Prolamins
  - Histones
  - Globins
  - Protamines
- Scleroproteins
  - Collagens
  - Elastins
  - Keratins
- Conjugated proteins
  - Nucleoproteins
  - Glycoproteins
  - Mucoproteins
  - Lipoproteins
  - Phosphoproteins
  - Chromoproteins
  - Metaloproteins
- Derived proteins
  - Primary proteins
  - Coagulated proteins
  - Proteans
  - Metaproteins
- Secondary proteins
  - Proteoses

- Peptones
- Polypeptides
- Peptides

LIMITING AMINO ACIDS

While protein synthesis take place those amino acid exzost first that is known as the limiting amino acid.

- Lysine and threonine are limiting amino acids in cereals and millets.
- Methionine and tryptophan are limiting amino acids in legumes.
- Methionine, lysine and threonine are limiting amino acids in peanuts.
- Methionine is a limiting amino acid in rice, soybean, and green leafy vegetables.
- Methionine and cysteine are limiting amino acids in beef protein.
- Lysine is a limiting amino acid in wheat gliadin and sunflower seed.
- Tryptophan is a limiting amino acid in maize.

BODY MASS INDEX

Most of the physiological, biochemical and body compositional changes associated with malnutrition occur in all its varieties and the principles of classification, investigation and management are the same for adults and children. Studying the body mass index, malnutrition can be classified into different five groups according their body mass index values.

$$BMI = Weight \ (kg)/[Height \ (m)]^2$$

Classification of Malnutrition in Adults by Body Mass Index (BMI).

Weight/height2 (kg/m^2)	Classification
20	Normal
18.5 - 20	Marginal
17 - 18.5	Mild malnutrition
16 - 17	Moderate malnutrition
16	Severe malnutrition

ENERGY AND PROTEIN REQUIREMENTS

For the growth of the body as well as for its maintenance, the essential nutrients required are proteins, carbohydrate, fats, minerals, vitamins and water.

Energy Allowance Required for Children and Adults

Category	Age group (years)	Body weight (kg)	Energy allowance (Kcal)
Children	4 - 6	18.87	1720
Boys	10 - 12	34.30	2420
	16 - 18	56.50	2820
Girls	10 - 12	36.47	2260
	16 - 18	50.00	2200
Man (moderate work)		55.00	2800
Woman (moderate work)		45.00	2200

Protein Requirement for Infants, Pre-school Children and Adults

Category	Age group (M/Y)	Body weight (Kg)	Protein required (g/kg/day)
Infants	6 - 7 M	-	1.80
Pre-school Children	4 - 6 Y	18.87	1.56
Boys	10 - 12 Y	34.30	1.24
Girls	10 -12 Y	36.47	1.17
Man		55.00	1.00
Woman		45.00	1.00

Ideal protein requirements per kg body weight per day are for infants: 1.85 g, pre school children 0.90 g, and adults 0.55 g.

Recommended Dietary Vitamin Allowances

Vitamin	Men	Women	Children up to age 11 years
Fat soluble			
Vitamin A (retinol, µg)	1000	800	400-700
Vitamin D (cholecalciferol, µg)	5-10	5-10	10
Vitamin E (α-tocopherol, mg)	10	8	6-7
Vitamin K (µg)	45-80	45-65	15-30
Water soluble			
Vitamin C (mg)	60	60	40-45
Vitamin B_1 (thiamin, mg)	1.5	1.1	0.7-1.0
Vitamin B_2 (riboflavin, mg)	1.7	1.3	0.8-1.2
Niacin (mg)	19	15	9-13
Vitamin B_6 (pyridoxine, mg)	2.0	1.6	1.0-1.4
Vitamin B_{12} (µg)	2.0	2.0	0.7-1.4
Folic acid (µg)	200	180	50-100

Approximate cost of 1 kg of Protein from Different Sources.

Protein source	Cost (Rs.)
Full-fat-soy-flour	50
Split pulses (Dhal)	100
Egg	250
Milk	300
Chicken	360
Fish	420
Meat	480

DEFINITIONS OR TERMS USED IN ASSESSMENT OF PROTEIN QUALITY

- Crude protein (CP) = Nitrogen % x 6.25
- True digestibility (TD) = [N absorbed ÷ N intake] × 100
- Biological value (BV) = [N retained ÷ N absorbed] × 100
- Net protein utilization (NPU) = [TD x BV] ÷ 100
- Utilizable protein (UP) = [NPU × Protein %] ÷ 100
- Nitrogen balance (NB) = Dietary N intake – Total N excretion
- Thomas-Mitchell method, BV = [Intake – output corrected for endogenous loss] [Intake - fecal losses corrected for endogenous loss]. OR

  = [Food N - (urine N - endogenous urine N) - (fecal N - metabolic N) [Food N - (Fecal N - metabolic N)].
- Apparent digestibility (AD) = [Food N - Fecal N] ÷ [Food N] x 100.
- Net protein ratio (NPR) = [Gain in wt. + Loss of wt. of non-protein group of animals] [Protein eaten].
- Net protein value (NPV) = BV x TD x % protein in food.
- Protein retention efficiency (PRE) = NPR x % protein in body.
- Protein efficiency ratio (PER) = Gain in weight ÷ Protein eaten.
- Chemical Score (CS) = Limiting amino acid in test protein expressed as a % age of the same amino acid in egg protein.
- Protein score (PS) = Limiting amino acid in test protein expressed as % age of the same amino acid in the reference amino acid pattern.
- Available lysine value (ALV) = Colorimetric test (by flurodinitro benzoate) of free amino groups in the lysine of the protein.
- Egg replacement value (ERV) = [100 - (N balance in egg - N balance in test protein)] ÷ [Food N].
- Gross protein value (GPV) = Extra growth produced by 3% protein supplement added to 8% cereal diet.

- Nitrogen balance Index (NBI) = Slope of line derived from plotting N balance against N intake.
- NPU standardized = NPU measured at that protein level which just maintains body weight (usually 4%).
- Net dietary protein energy ratio (NDPE %) = NPUop × protein energy expressed as ratio of total energy.
- Net dietary protein value (NDPV) = NPUop × g_2 protein index.
- Protein rating = PER × protein % × reasonable daily intake of food.

Multification Factor while Calculating Protein Content From Different Food and Food Products

Food material	Multification factor
Sorghum, Bajara, other millets and Maize	6.25
Rice	5.95
Whole wheat	5.83
Wheat bran	6.31
Wheat flour (free from germ or bran)	5.70
Bulgur wheat	5.83
Rye	5.83
Barley, Oats	5.83
Groundnut (peanuts) and Brazil nuts	5.46
Soybean	5.71
Sesamum	5.30
Sunflower	5.36
Almonds	5.18
Other nuts and oilseeds	5.30
Casein and other milk proteins	6.38

Nutritional Value of Some Protein Foods.

Food	Chemical Score [based on FAO ref]	PER	BV	NPU
Human milk	100	4	95	87
Cow's milk	94	3.1	84	82
Hen's egg	100	3.9	94	94
Meat (beef)	100	3.0	74	67
Fish all type	100	3.5	76	79
Wheat grain	56	1.5	65	40
Rice brown	73	2.0	73	70
Soybeans	80	2.3	73	61

Synergistic Effects of Pulse Supplementation to Cereal Diets.

Diet	Protein value expressed as %		
	TD	BV	NPU
Wheat chapati	89	66	59
Soybean	70	52	36
Wheat + Soybean chapatti	84	79	66
Wheat (raw) (3:1)	84	61	51
French bean (autoclaved)	68	58	39
Wheat + French bean (autoclaved (7:3)	81	81	65

Iso-electric Points of Some Food Proteins.

Protein	Iso-electric point (pH)
Casein	4.55
Egg albumin	4.55 - 4.90
Gelatin	4.80 - 4.85
-Lacto globulin	5.2
Myosin	6.2 - 6.6
Gliadin	6.5

Functional Properties of Proteins Required in Various Foods.

Food	Functionality
• Beverages	Solubility at different pH, heat stability, viscosity.
• Soups, sauces	Viscosity, emulsification, water retention.
• Dough formation, baked products, e.g., Bread, cakes etc.	Formation of a matrix and film with viscoelastic properties, cohesion, heat denaturation, gelation, water absorption, emulsification, foaming, browning.
• Dairy products, e.g., processed cheese, ice cream, desserts etc.	Emulsification, fat retention, viscosity, foaming, gelation, coagulation.
• Egg products, e.g., substitutes	Foaming, gelation.
• Meat products, e.g., sausage	Emulsification, gelation, cohesion, water and fat absorption and retention.
• Meat extenders, e.g., texturized vegetable proteins.	Water and fat absorption and retention, insolubility, hardness, chewiness, cohesion, heat denaturation.
• Food coatings	Cohesion, adhesion.
• Confectionary products, e.g., milk chocolate	Dispersibility, emulsification

Causes and Effect in the Deterioration of Food.

Some primary events	Consequence	Quality change
Hydrolysis of lipids	Free fatty acids react with protein	Texture, flavour, nutritive value
Hydrolysis of polysaccharides	Sugars react with proteins	Texture, flavour, colour, nutritive value
Oxidation of lipids	Oxidation products reacts with many other constituents	Texture, flavour, colour, nutritive value
Bruising of fruit	Cells break, enzymes are released, oxygen accessible	Texture, flavour, colour, nutritive value
Heating of green vegetables	Cell walls and membranes lose integrity, acids and enzymes are released	Texture, flavour, colour, nutritive value
Heating of muscle tissue	Proteins denature & aggregate, enzymes become inactive	Texture, flavour, colour, nutritive value

Many reactions can lead to the deterioration of food quality or impairment of food safety.

Classification of Undesirable Changes that can Occur in Food.

Attribute	Undesirable changes
Texture	Loss of solubility
	Loss of water holding capacity
	Toughening
	Softening
Flavour	Development of
	Rancidity (hydrolytic or oxidative)
	Cooked or caramel flavours
	Other off-flavours
Colour	Darkening
	Bleaching
	Development of other off colours
Nutritive value	Loss or degradation of
	Vitamins
	Minerals
	Proteins
	Lipids
	Carbohydrates

BROWNING REACTIONS

Browning of foods is due to oxidative or nonoxidative reactions. Oxidative, or enzymatic browning, is a reaction between oxygen and a phenolic substrate catalyzed by polyphenol oxidase. This is the common browning that occurs in cut apples, bananas, pears, and even lettuce, and it does not involve carbohydrates. Non-oxidative or non-enzymatic browning is of widespread importance in foods. It involves the phenomena of carmelization and/or the interaction of proteins or amines with carbohydrates. The latter reaction is the Millard reaction. Direct heating of carbohydrates, particularly sugars syrups, produces a complex group of reactions termed "Caramelization". Reaction is facilitated by small amounts of acids and certain salts.

Sucrose is commonly used for making caramel colours and flavours. The most abundant is acid-fast caramel made with ammonium bisulfite catalyst to produce the colour for cola drinks. Another is a brewer's colour for beer, made by heating a sucrose solution with ammonium ion, and the third is a baker's colour produced by direct pyrolysis of sucrose to give a burnt sugar colour.

DEHYDRATION AND THERMAL DEGRADATION OF CARBOHYDRATES

Dehydration and thermal degradation of sugars are important reactions in foods and are catalyzed by acid or base and many of the β-elimination type. Pentoses yield 2-furaldehyde as the main dehydration product; hexoses yield 5-hydroxymethyl-2-furaldehyde (HMF) and other products, such as 2-hydroxyacetylfuran and isomaltol. Fragmentation of the carbon chain of these primary dehydration products leads to other chemical species, such as levulinic acid, formic acid, acetol, acetoin, diacetyl, lactic acid, pyruvic acid, and acetic acid. Some of these degradation products are highly odorous and may have desirable or undesirable flavours.

WHAT ARE THE BIOCHEMICAL FUNCTIONS OF CARBOHYDRATES?

Source of energy

Conversion of solar energy in to chemical energy

Central to metabolism

Utilization of fats

Sweetness of food

Appetizing and flavouring

Bulk of food

Peristaltatic movement of stomach

Movement of joints

Structural support

Storage of food material

Transport between cells and organs

Accumulation of fats

Helps in pollination

Helps in transamination

Helps for fat burns

Responsible for ketone body formation

It exerts sparing effect of protein

It acts as drugs

It helps in protein synthesis

Responsible for colour development to the products

The ability to bind water and control water activity in foods is one of the most important properties of carbohydrates

It helps in the sugar-flavourant formation

A burnt carbohydrate gives colour to the food products

It also acts as a lubricant at joints

WHAT IS THE GLUCOSIDIC BONDING BETWEEN VARIOUS SUGARS?

Maltose	$\alpha(1, 4)$	Cellulose	$\beta(1, 4)$
Isomaltose	$\alpha(1, 6)$	Gentibiose	$\beta(1, 6)$
Sucrose	$\alpha, 1, \beta, 2$	Lactose	$\beta(1, 4)$
Lactulose	$\alpha(1, 4)$	Melibiose	$\alpha(1, 6)$
Raffinose	$\alpha(1, 2)\ \alpha(1, 6)$		
Sachyose	$\alpha(1, 2)\ \alpha(1, 6)\ \alpha(1, 6)$		
Verbascose	$\alpha(1, 2)\ \alpha(1, 6)\ \alpha(1, 6)\ \alpha(1, 6)$		
Amylose	$\alpha(1, 4)$		
Amylopectin	$\alpha(1, 4)\ \alpha(1, 6)$		

The pH Range of Common Indicators

Name	PH range	Colour in acid solution	Colour in alkaline solution
Acid cresol red	0.2 - 1.8	Red	Yellow
Acid meta cresol purple	1.2 - 2.8	Red	Yellow
Thymol blue	1.2 - 2.8	Red	Yellow
Benzo yellow	2.4 - 4.0	Red	Yellow
Metyl orange	2.9 - 4.6	Red	Orange/Yellow
Bromocresol green	3.8 - 5.4	Yellow	Blue
Bromophenol blue	3.0 - 4.6	Yellow	Blue
Methyl red	4.4 - 6.0	Red	Yellow
Chlorophenol red	5.2 - 6.8	Yellow	Red
Bromocresol purple	5.2 - 6.8	Yellow	Purple
Bromothymol blue	6.0 - 7.6	Yellow	Blue
Phenol red	6.8 - 8.4	Yellow	Red
Cresol red	7.2 - 8.8	Yellow	Red
Metacresol purple	7.6 - 9.2	Yellow	Purple
Thymol blue	8.0 - 9.6	Yellow	Blue
Phthalein red	8.6 - 10.2	Yellow	Red
Toyl red	8.6 - 11.6	Red	Yellow
Parazo orange	11.0 - 12.6	Yellow	Orange
Acyl blue	12.0 - 13.6	Yellow	Blue

PROBLEMS IN JELLY FORMATION

Pectin is the most important constituent of jelly. It is possible to achieve a desired level of consistency by utilizing proper proportion of pectin, sugar and acid. Pectin 1%, T.S.S. 60%, pH 3.0 to 3.5 and small quantity of calcium salt are needed for obtaining firmness to the jelly.

- Deformities in jelly
- Failure of jelly to set
  - Lack of acid or pectin
  - Addition of too much sugar
  - Cooking below end point
  - Cooking beyond end point
  - Slow cooking for a long time

- Cloudy jelly
  - Over cooking
  - Faulty pouring
  - Non-removal of scum
  - Premature gelation
  - Use of non-clarified juice
  - Use of immature fruits
  - Over cooking
- Synersis
  - Excess of acid
  - Too low concentration of sugar
  - Insufficient pectin
  - Premature gelation
  - Fermented jellies

CLOUDY OR FOGGY JELLIES

Cloudy jelly : This type of jelly is formed if the juice or extract not clarified. Use of immature fruits, green, immature fruits contain starch, which is insoluble in the juice, and the jelly made from them has the cloudy appearance.

Over-cooling : When the jelly is cooled too much, it becomes viscous and sometimes lumpy. Such a jelly is almost cloudy.

Faulty pouring into containers : The jelly should not be poured in containers from a great height because air gets incorporated into the mass and the bubbles formed do not clear easily, especially when the jelly is well made and sets within a short period. The spout of the pouring vessel should not be more than about 2.5 cm from the top of the container.

Non-removal of scum : The jelly becomes cloudy also when scum is not removed before pouring.

Premature gelation : If there is excess of pectin in the juice, it causes premature gelation with the result that air may get trapped and make jelly opaque. This can be avoided by using low pectin.

Formation of crystals : Adding excess sugar causes the formation of sugar crystals in a jelly. It is also an indication of over-concentration of the jelly.

Syneresis or weeping : The phenomenon of spontaneous exudation of fluid from a gel is syneresis or weeping of jelly. It is caused by several factors such as excess of acid, too low concentration of sugar or soluble solids and insufficient pectin.

Temperature Tests for Syrups and Candies

Product	Temperature (°C)	Test	Description of test
Syrup	110-112	Thread	When syrup is dropped from a spoon, syrup spins 5 cm thread
Barfi, fondant fudge	112-115	Soft ball	Forms a soft ball when syrup is dropped into cold water, flattens on removal from walls
Caramels	118-120	Firm ball	Forms a firm ball when syrup is dropped into cold water, does not flatten on removal from water
Divinity, laddu, marsh mellow	120-130	Hard ball	Forms a hard ball enough to hold its shape when syrup dropped in cold water
Butter scotch, toffees	132-143	Soft crack	Forms threads, which are hard but not brittle when syrup dropped in cold water
Brittle, glace	150-154	Hard crack	Forms threads, which are brittle when syrup is, dropped sugar melts
Barley sugar	160	Clear liquid	Sugar melts
Caramel	170	Brown liquid	Sugar melts and browns

WINE

Wine is a fermented product of fruit pulp or juice. For fermentation of pulp or juice Saccharomyces cerevisiae var. ellipsoideus yeast is used.

Fruit Wines are Classified as Table Wine or Dessert Wine.

Wine	Sugar content (g/100 ml)	Alcohol content (ml/100 ml
Table purpose		
Dry	0 - 1	9 - 11
Semi dry	2 - 3	10 - 12
Dessert		
Slightly sweet	3 - 8	11 - 13
Sweet	8 - 11	12 - 14
Very sweet	11	13 - 18

Alcoholic Drink and Their Calorie Values.

Type	Alcohol content (%)	Calorie value (Cal/100 ml)
Beer		
Bitter	3	30
Mild	3	25

Contd...

Contd...

Type	Alcohol content (%)	Calorie value (Cal/100 ml)
Cider		
Dry	4	35
Sweet	4	40
Wine		
White	9	70
Red	9	65
Fortified	16	135
Spirit		
Whisky	31	220
Liquor		
Benedictine	39	270
Port	21	293
Gin40	210	
Rum (sugarcane)	40	210
Mixers		
Regular	0	135
Sugar free	0	0

Different Unfermented and Fermented Fruit Beverages

Unfermented beverages	Fermented beverages
Juice	Wine (7-9% alcohol light, 9-16% alcohol medium, 16-21% alcohol strong)
RTS	Champaigne (sparkling wine)
Nectar (20% fruit pulp, 3% acidity)	Port (Portugal)
Cordial (sparkling, 25% juice, 30% TSS, 1.5% acidity)	Sherry (Spanish wine)
Squash (30% juice, 45% sugars)	Tokay (Fortified wine, Hungary)
Crush (25% juice, 55% TSS, 1% acidity)	Muscat (Muscat grapes)
Syrup (40 - 45% juice, 59% sugars)	Perry (Prepared from pears)
Fruit juice concentrate	Orange wine (From oranges)
Fruit juice powder	Berry wine (From various types of berries)
Barley water (25% juice, 30% TSS, barley starch 0.25%, 1% acidity)	Nira (Palm tree juice)
Carbonated beverage	Feni (Cashew apple, Goa)
	Cider (From apple)

PRESERVATION OF FOOD

Food and food product preservation methods are based on basic six principles.

1. **Removal of moisture**
   - Drying
   - Dehydration
   - Concentration
   - Intermediate moisture processing
   - Freeze-drying

   Removal of the biologically active water through drying or dehydration stops the growth of microorganisms.

2. **Heat treatment**

   Heat treatment results in denaturation of proteins, i.e., inactivation of the microbial and native food enzymes.
   - Blanching
   - Pasteurization
   - Sterilization

3. **Low-temperature treatment**

   Low temperature inhibits microbial growth and slows down the rate of chemical and enzymatic reactions.
   - Cold storage
   - Refrigeration
   - Freezing
   - Dehydro-freezing

4. **Acidity control**

   The growth of food spoilage organism is significantly inhibited in an acidic environment.
   - Fermentation
   - Acidic additive

5. **Use of preservatives**

   Chemical additives can substantially contribute to the preservation of foods by providing inhibitory environment for microbial growth and for enzymatic and chemical reactions.
   - Sodium benzoate

- Potassium metabifulfite
- Benzoic acid
- Sugars and salts

6. Irradiation

The free radical mechanism of irradiation destroys microorganisms but is also detrimental to nutrients particularly vitamins.

- Gamma rays
- Cobalt-60

Preservation Methods

Type	Product to be preserved
Physical methods of preservation	
• By removal of heat	Refrigeration, freezing preservation, dehydration,
(preservation by cold)	dehydro freezing preservation, carbonation
• By addition of heat (thermal processing)	Stationary pasteurization, agitating pasteurization or sterilization, flash pasteurization or HTST processing etc.
• By removal of water (evaporation or dehydration)	Sun-drying, dehydration, low temperature evaporation or concentration, freeze-drying, Accelerated freeze-drying, foam-mat drying, puff drying etc.
• By irradiation	Dosing with UV or ionizing radiation etc.
Chemical methods of preservation	
• By addition of acid such as vinegar or lactic acid	Pickled vegetables, fish and meat
• By salting or brining	Vegetable or fruit pickles, salted fish, salt-cured meat and pork etc.
• By addition of sugar and heating	Fruit preserves, jams, jellies, marmalades etc.
• By addition of chemical preservatives	Using water-soluble salts of sulphur-dioxide benzoic acid, sorbic acid and a few like hydrogen peroxide etc., which are permitted as harmless in foods. By means of substances of bacterial origin such as tylosin, resin etc. which are permitted to a limited extent, in some cases as harmless additives.
• By fermentation	Alcoholic and acetous fermentation as in the case of fruit wines, apple cider, fruit, vinegar etc.

CLASSIFICATION OF MAJOR FOOD PRESERVATION TECHNIQUES

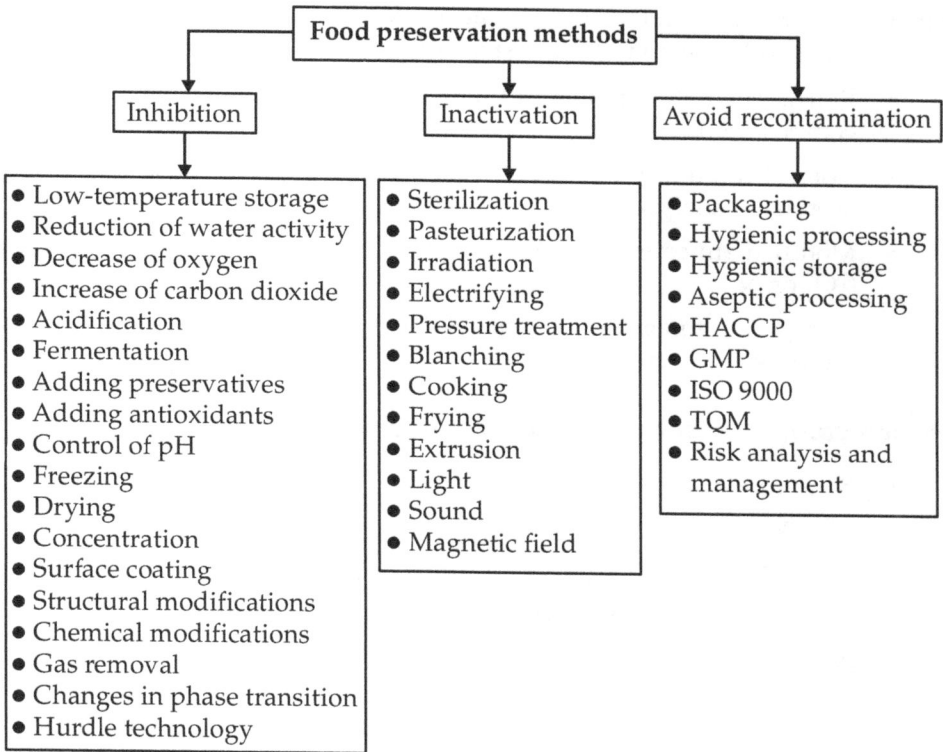

```
                    ┌─────────────────────────────┐
                    │   Food preservation methods  │
                    └─────────────────────────────┘
        ┌────────────────┬────────────────────┬────────────────────────┐
        ▼                        ▼                         ▼
  ┌──────────┐          ┌──────────────┐        ┌───────────────────────┐
  │Inhibition│          │ Inactivation │        │ Avoid recontamination │
  └──────────┘          └──────────────┘        └───────────────────────┘
```

Inhibition	Inactivation	Avoid recontamination
• Low-temperature storage • Reduction of water activity • Decrease of oxygen • Increase of carbon dioxide • Acidification • Fermentation • Adding preservatives • Adding antioxidants • Control of pH • Freezing • Drying • Concentration • Surface coating • Structural modifications • Chemical modifications • Gas removal • Changes in phase transition • Hurdle technology	• Sterilization • Pasteurization • Irradiation • Electrifying • Pressure treatment • Blanching • Cooking • Frying • Extrusion • Light • Sound • Magnetic field	• Packaging • Hygienic processing • Hygienic storage • Aseptic processing • HACCP • GMP • ISO 9000 • TQM • Risk analysis and management

Precooling

Precooling is a means of removing field heat. The aim is to quickly slow down respiration of produce, minimize microbial growth, reduce transpiration rate and ease the load on cooling system.

- Air-cooling

  Forced-air (1.7°C) cooling may last about 2-4 hours.

- Hydro cooling

  Cold water or ice may be used for lowering field heat from the produce to increase the shelf life.

- Vacuum cooling

  This is very rapid method of precooling mostly used for leafy vegetables.

Storage

Food commodity often requires some storage to balance day-to-day fluctuations between harvest and sale or for long-term storage to extend marketing beyond the end of harvest season.

- Refrigerated storage (control temperature, humidity and atmospheric composition)
- Controlled atmosphere (CA) storage (use of CO_2, O_2, CO, C_2, H_4, N_2 or acetylene manipulation)
- Modified atmosphere (MA) storage (Inter changeable atmosphere with CO_2 or N_2)
- Hypo baric storage: It is a type of CA storage with emphasis on reducing the pressure exerted on storage material.

Which components come under cooling and storage?

- Background
    - Temperature
    - Relative humidity
    - Atmospheric composition
- Pre-cooling
    - Hydro cooling
    - Contact icing
    - Vacuum cooling
    - Forced-air cooling
- Cooling and refrigeration
    - Room cooling
    - Mechanical refrigeration
    - Evaporative cooling
    - Carbon dioxide cooling
    - Alternative methods
        - Nighttime cooling
        - Well water-cooling
        - High-altitude cooling
        - Underground storage
        - Thermoelectric cooling (peltier effect)

- Modification of atmospheric composition
  - Controlled atmosphere storage
  - Oxygen control systems
  - Carbon dioxide control systems
  - Ethylene control systems
  - Modified atmosphere storage

What are the packaging materials uses for storage of different kinds of food items?

1. Tin cans:

   Electrolytic tinning

   Steel base Type L, MR, MC

   Tin coating

2. Glass:

   White; glass containers, bottles, jars, tumblers etc.

   Coloured; jugs, carboys, vials and ampules etc.

3. Plastic and films
4. Papers
5. Laminations
6. Solid and corrugated board containers
7. Wooden boxes and crates
8. Semi-rigid packaging materials and packaging forms
9. Films; amylase film
10. Ionomers
11. Aluminium foil
12. Flexible packaging materials

CANNING MATERIAL

1. Metal containers
   - Acid resistant cans
   - Sulphur resistant cans
2. Glass containers

Thick superior quality glass containers with improved twist off lids are required. Most of the glass material is inert and responsible to avoid any chemical reaction on storage container.

SPOILAGE OF CANNED FOODS

Spoilage occurs due to physical, chemical or by microbial changes.

1. Symptoms of spoilage
   - Swelling or bulging of cans
   - Hydrogen swell
   - Flipper
   - Springer
   - Flat sour
   - Leaker
   - Bursting of cans
   - Breather

2. Discolouration of products
   - Chemical reactions
   - Metallic contamination
   - Ferric tannate
   - Iron sulphide
   - Copper sulphide
   - Hydrogen swelling
   - Corrosion due to oxygen

3. Microbial spoilage
   - Non-poisonous spoilage
   - Poisonous spoilage
   - Spoilage by fungi

CLIMACTERIC AND NON-CLIMACTERIC FRUITS

No clear metabolic difference has been demonstrated between climacteric and non-climacteric fruits, although ripening in fruits categorized as non-climacteric often proceeds more slowly.

Classification of Fruits According to Respiratory Patterns.

Climacteric fruit	Non-climacteric fruit
Apple	Blueberry
Apricot	Cacao
Avocado	Cherry
Banana	Cucumber
Cherimoya	Fig
Custard apple	Grape
Feijoa	Grape fruit
Mango	Lemon
Papaya	Litchi
Passion fruit	Melon
Papaw	Olive
Peach	Orange
Pear	Pineapple
Plum	Pomegranate
Sapota	Rin-tomato
Tomato	Strawberry
Watermelon	Tamarillo
Irregularity in the rate of respiration	Regularity in the rate of respiration
Maximum rate of respiration at ripeness and then decline	Non-climacteric fruits generally exhibits a decline in respiratory activity during storage

*Vegetables do not exhibit the climacteric phenomenon.

FLAVOUR

Flavour is the subtle and complex sensation that is the source of much of the delight man finds in food. Flavour is a combination of taste, smell and feel. There are many taste buds in mouth capable of detecting sweet, sour, salty and bitter. Taste sense is detected through the solution of soluble compounds in the saliva or in the food juices and the contact of those dissolved compounds with the taste buds.

Synthetic Flavouring Substances

Aromatic chemical	Flavour	Average use (ppm)
Group-I		
• Vanillin	Vanilla	31.5
• Ethyl vanillin	Vanilla	16.6
• Citral	Lemon	17.6

Contd...

Aromatic chemical	Flavour	Average use (ppm)
• Benzaldehyde	Cherry, almond	84.8
• Cinnamic aldehyde	Cinnamon, cola	110.7
• Methyl salicylate	Root beer, wintergreen	129.7
• Amyl acetate	Banana fruit flavours	78.4
• Amyl butyrate	Banana fruit flavours	23.7
• Ethyl acetate	Fruit, rum	92.7
• Ethyl butyrate	Strawberry, fruit	31.1
• Methyl anthranilate	Grape	97.1
• Ethyl oxyhydrate	Rum	472.7
• Anethole	Root beer, anise	133.5
• Safrot	Root beer, anise	16.9
• Methanol	Mint	111.2
Group-II		
• Heliotropine	Vanilla, cherry	3.4
• Aldehyde C-18	Coconut	17.6
• Diacetyl	Butter	17.3
• Aldehyde C-16	Strawberry	10.0
• Ethyl oenanthate	Grape	8.1
• Allyl caproate	Pineapple	11.7
• Aldehyde C-14	Peach	8.0
• Eugenol	Spice	48.8
• Amyl valerinate	Peach, fruit	9.6
• Ethyl valerinate	Fruit	14.4
• Carvone	Spice, mint	190.3
• Anis aldehyde	Cherry, Vanilla	3.6
Group-III		
• Benzyl acetate	Straw berry, fruit	8.8
• Ethyl formate	Rum, Straw berry	77.0
• Ionone	Raspberry	0.9
• Ethyl lactate	Grape	42.1
• Toyl aldehyde	Cherry	9.3
• Amyl caproate	Pineapple, apple	4.4
• Linalool	Spice	30.2
• Citronellal	Rose (general)	14.2

Contd...

Aromatic chemical	Flavour	Average use (ppm)
• Butyl acetate	Fruit	26.4
• Ethyl pelargonate	Grape, fruit	2.3
• Butyl butyrate	Fruit	17.9
• Aldehyde C-10	Citrus	1.2
• Ethyl heptoate	Grape, pineapple	4.6
• Geranyl acetate	Fruit	8.4
• Methyl phenyl acetate	Honey	2.4
Group-IV		
• Aldehyde C-8	Citrus	0.71
• Aldehyde C-9	Citrus	2.81
• Cyclohexyl butyrate	Pineapple, fruit	8.3
• Ethyl laurate	Spice, fruit	6.9
• Linalyl acetate	Citrus	5.4

PRESERVATIVES

Preservatives mean a substance which when added to food is capable of inhibiting, retarding or arresting the process of fermentation, acidification or other decomposition of food.

Class-I preservative	Class-II preservative
Common salt	Benzoic acid and their salts
Sugar	Sulphurous acid and their salts
Dextrose	Nitrites of sodium or potassium
Glucose (syrup)	Sorbic acid including its sodium, potassium and calcium salts
Spices	Nisin, Sodium, potassium and calcium salts of lactic acid
Vinegar or acetic acid	Sodium and calcium propionate
Honey	Methyl or propyl parahydroxy-benzoate
Edible vegetable oil	Propionic acid and its esters or salts
* Addition of these preservatives in any food is not restricted	* Addition more than one and exceeding the given limit is prohibited

Food Colours and Their Uses in Fruit and Vegetable Products.

Colour	Products
Black currant A	Black current jam and jelly
Carmoisine WNS	Raspberry jelly
Caramel W	Vinegar, Sauces
Erythrosin AS	Canned cherries, Victoria plums
Orange AG	Orange jelly, orange squash
Pea Green B	Canned fresh and processed peas
Ponceau RNS	Canned straw berries, strawberry jam and jelly
Red BR	Canned beet root
Sunset Yellow FCF	Apricot jam
Tartrazine NSS	Lemon jelly, lime juice cordial

Use of Emulsifier in Food Industry.

Emulsifier	Applications in food industry
Monoglyceride (Glyceryl monostearate-GMS)	Margarine, ice cream, bread, potato products, snack foods
Diacetyl tartaric acid ester of mono-glycerides (DATE)	Tomato juice, tea concentrate, orange juice, in baking industry as a dough conditioning agent
Succinylated mon-glycerides	In baking industry as a dough conditioner and bread softener
Calcium stearoyl lactoyl lactate (WL)	Whipping agents in egg whites and dehydrated potato
Sodium stearol lactoyl lactate (SSL)	In baking industry as dough conditioners and antistaling agents
Span 60 (sorbitan monostearate), Span 80 (sorbitan monoleate),	Tween (poly-oxyethylene sorbitan monostearate) To impart desirable structural characteristics to the frozen ice cream, i.e. stand-up properties in frozen ice cream to enable it to be extended into shapes.
Polyglycerol esters propylene glycol monoesters	These emulsifiers (also monoglycerides, SSL. Polyoxyethylene sorbitan esters) are used in baking fats or as individual additives to improve cake quality.

Permissible Levels of SO$_2$ Preservatives in Processed Fruits and Vegetables Products.

Product	Preservative	Permissible level (ppm)
Alcoholic wines	Sulphur dioxide	450
Cherries, strawberries and raspberries, fruit pulp, jam, juices	Sulphur dioxide	2000

Contd...

Product	Preservative	Permissible level (ppm)
Canned cherry and fruit jelly	Sulphur dioxide or Benzoic acid	40
Cured fruit	Sulphur dioxide	150
Cider	Sulphur dioxide	200
Dried fruits	Sulphur dioxide	2000
Dehydrated vegetables	Sulphur dioxide	2000
Fruit juice concentrates	Sulphur dioxide	1500
Fruits and vegetable flakes, powder, figs	Sulphur dioxide	600
Jam, marmalade, preserve	Sulphur dioxide	40
Jaggery, cider	Sulphur dioxide	200
Non-alcoholic beverages, wines, squashes, crushes, fruit syrups, cordials, fruit juices	Sulphur dioxide	350
Pickles and chutneys	Benzoic acid or Sulphurdioxide	250 or 100
Raisins and sultanas	Sulphur dioxide	750
Syrups and sherbets	Sulphur dioxide or Benzoic acid	350
Tomato and other sauces	Benzoic acid	750
Tomato puree and paste	Benzoic acid	750
White sugar, dextrose	Sulphur dioxide	70

Preservatives Permitted in Food Items.

Preservative	Concentration (ppm)	Foods
Sulphur dioxide or salts of sulphurous acid	50 - 3000	Sausages, fruit pulps, fruit juice conc., Squashes, crushes, jams, syrups, cordials, sugar glucose, khandsari, corn flour syrup, beer, cider, pickles, dehydrated vegetables, wines, ready-to-serve beverage, hard boiled confectionery.
Propionic acid and its salts	5000	Bread
Benzoic acid and its salts	50 - 6000	Syrups, squashes, jams, jellies, RTS, pickles, chutney, ginger beer, tinned caviare.
Sorbic acid and its salts	1000 - 1500	Cheese, bread flour, confectionery, smoked fish, wrappers.
Nisin	1000	Cheese
Nitrites	200	Cooked pickled meat, ham and bacon
Nitrates	500	Cooked pickled meat, ham and becon

Commonly Used Food Additives and Their Functions.

Class and general function	Chemical name of food additive
A. Processing additive	
• Aerating and foaming agents	Carbon dioxide, Nitrogen, Sodium bicarbonate
• Antifoam agents	Aluminium stearate, Ammonium stearate, Butyl stearate, Decanoic acid, Dimethyl Ploysiloxane, Dimethylpoly-silicone, Lauric acid, Mineral oil, Oleic acid, Oxystearin, Palmitic acid, Petroleum waxes, Silicon dioxide, Stearic acid.
• Catalysts (including enzymes)	Nickel, Amylase, Glucose oxidase, Lipase, Papain, Pepsin, Rennin.
• Clarifying and flocculating agents	Bentonite, Gelatin, Polyvinyl pyrrolidone, Tannic acid.
• Colour control agents	Ferrous gluconates, Magnesium chloride, Nitrate, Nitrite (potassium, sodium), Sodium erythrobate
• Freezing and cooling agents	Carbon dioxide, Liquid nitrogen, Freezant-112, Cl_2, CF_2
• Malting and Fermenting aids	Ammonium chloride, Ammonium sulfate, Ammonium phosphate (dibasic), calcium carbonate, Calcium phosphate, Calcium sulfate, Potassium chloride, Potassium phosphate (dibasic).
• Material handling aids	Aluminum phosphate, calcium silicate, Calcium stearate, Dicalcium phosphate, Dimagesium phosphate, Kaolin, Magnesium silicate, Starches, Tricalcium phosphate, Tricalcium silicate, Xanthan and other gums.
• Oxidizing-reducing agents	Acetone peroxides, Benzoyl peroxide, Calcium peroxide, Hydrogen peroxide, Sulfur dioxide.
• pH control and modification agents (acidulants, acids)	Acetic acid, Citric acid, Fumaric acid, Lactic acid, δ-Gluconolactone, Malic acid, Hydrochloric acid, Tartaric acid, Phosphoric acid, Succinic acid, Potassium acid tartrate.
• Alkalies (bases)	Ammonium bicarbonate, Ammonium hydroxide, Calcium carbonate, Magnesium carbonate, Potassium bicarbonate, Potassium hydroxide, Sodium bicarbonate, Sodium carbonate, Sodium citrate, Trisodium phosphate
• Buffering agents	Ammonium phosphate, Calcium citrate, Calcium gluconate, Calcium phosphate, Potassium acid tartrate, Potassium citrate, Potassium phosphate, Sodium acetate, Sodium acid pyrophosphate, Sodium citrate, Sodium phosphate, Sodium potassium tartrate.
• Release and antistick agents	Acylated monoacylglucerols, Bees wax, Calcium stearate, Magnesium silicate, Mineral oil, White mono- and Di-acylglycerols, Starches, Stearic acid, Talc.
• Sanitizing and fumigating agents	Chlorine, Methyl bromide, Sodium hypochlorite

Contd...

Contd...

Class and general function	Chemical name of food additive
• Separation and filtration aids	Diatomaceous earth, Ion-exchange resins, Magnesium silicate, Charcoal
• Solvents, Carriers and encapsulating agents	Acetone, Agar-agar, Arabinogalactan, Cellulose, Glycerine, Guar gum, Methylene chloride, Propylene glycol, Triethyl citrate
• Washing and surface removal agents	Sodium dodecyl benzene sulfonate, Sodium hydroxide

B. Final product additives

• Antimicrobial agents	Acetic acid (and salts), Benzoic acid (salts), Ethylene oxide, ρ-Hydroxy benzoate alkyl esters, Nitrates, Nitrites (Na, K, salts), Propionic acid, Propylene oxide, Sorbic acid, Sulfur dioxide and sulfites.
• Antioxidants	Ascorbic acid (and salts), Ascorbyl palmitate, BHA, BHT, Gum guaiac, Propylgallate, Sulfite and metabisulfite salts, Thiodipropionic acid (and salts).
• Appearance control agents (Colours and colour modifiers)	Annatto, Beet powder, Caramel, Carotene, Cochineal extract, FD and C Green No. 3, FD and C Red No. 3, Titanium dioxide, Turmeric, Beeswax, Glycerine, Oleic acid, Sucrose, Wax, Carnauba.
• Flavour and flavour modifiers (Flavouring agents)	Essential oils, Herbs and Spices, Plant extracts, Synthetic flavour compounds.
Flavour potentiators	Disodium guanylate, Disodium inosinate, Maltol, Monosodium glutamate, Sodium chloride
• Moisture control agents	Glycerin, Gum acacia, Invert sugar, propylene glycol, Mannitol, Sorbitol
• Nutrient, dietary supplements, amino acids	Alanine, Arginine, Aspartic acid, Cysteine, Cystine, Glutamic acid, Histidine, Isoleucine, Leucine, Lysine, Methionine, Phenylalanine, Proline, Serine, Threonine
• Minerals	Boric acid, Calcium carbonate, Calcium citrate, Calcium phosphate, Calcium pyrophosphate, Calcium sulfate, Cobalt carbonate, Cobalt chloride, Cobalt sulfate, Cupric chloride, Cupric gluconate, Cupric oxide, Cupric sulfate, Calcium fluoride, Ferric phosphate, Ferric pyrophosphate, Ferrous gluconate, Ferrous sulfate, Iodine, Iodide, Cuprous, Iodate, Potassium magnesium oxide, Magnesium phosphates, Magnesium sulfate, Magnesium chloride, Manganese citrate, Manganese oxide, Molybdate, Nickel sulfate, Phosphates, Calcium phosphates, Sodium potassium chloride, Zinc chloride
• Vitamins	ρ-Amino benzoic acid, Biotin, Carotene, Folic acid, Niacin, Niacin amine, Pantothenate, Calcium pyridoxine hydrochloride, Riboflavin, Thiamine hydrochloride, Tocopherol acetate, Vitamin A acetate, Vitamin B_{12}, Vitamin D.

Contd...

Contd...

Class and general function	Chemical name of food additive
• Miscellaneous nutrients	Betaine hydrochloride, Choline chloride, Inositol, Linoleic acid, Rutin
• Sequestrants (chelating agents)	Calcium citrate, Calcium disodium EDTA, Calcium gluconate, Calcium phosphate (monobasic), Citric acid, phosphoric acid, potassium citrate, Potassium phosphate, Sodium citrate, Sodium acid pyrophosphate, Sodium gluconate, Sodium hexametaphosphate, Sodium phosphate, Sodium potassium tartrate, Sodium tartrate, Sodium triphosphate, Tartaric acid.
• Surface tension control agents	Dioctyl sodium sulfosuccinate, Oxbile extract, Sodium phosphate (dibasic)
• Sweeteners (non-nutritive)	Ammonium saccharin, Calcium saccharin, Saccharin, Sodium saccharin
• Nutritive	Aspartame, Glucose, Sorbitol
• Texture and consistency control agents (emulsifiers and emulsifieracid, salts)	Calcium stearoyl-2-lactylate, Cholic acid, Deoxy cholic Dioctyl sodium sulfosuccinate, Fatty acids (C_{10} - C_{18}), Lecithin, Mono- and diacylglycerols, Oxbile extract, Polyglycerol esters, Potassium phosphate, Potassium pyrophosphate, Potassium polymetaphosphate, Sodium aluminum phosphate, Sodium citrate, Sodium phosphate, Sorbitan monosterate, Taurocholic acid (salts)
• Firming agents	Aluminum sulfate, Calcium carbonate, Calcium chloride, Calcium citrate, Calcium gluconate, Calcium hydroxide, Calcium lactate, Calcium phosphate, Calcium sulfate, Magnesium chloride
• Leavening agents	Ammonium bicarbonate, Ammonium phosphate, Calcium phosphate, Glucono-δ-lactone, Sodium acid pyrophosphate, sodium aluminum phosphate, Sodium aluminum sulfate, Sodium bicarbonate
• Masticatory substances	Paraffin (synthetic), Pentaerythritol ester of rosin
• Propellants	Carbon dioxide, Feron-115, Nitrous oxide
• Stabilizers and thickeners	Acacia gum, Alginic acid, Carrageenan, Guar gum, Hydroxypropylmethyl cellulose, Locust bean gum, Methyl cellulose, Pectin, Sodium carboxy methyl cellulose, Tragacanth gum
• Texturizers	Carrageenan, Mannitol, Pectin, Sodium caseinate, Sodium citrate
• Tracers	Titanium dioxide

Pigments Present in Food and Food Products

- Chlorophylls
- Myoglobin and hemoglobin

- Anthocyanins
- Flavonoids
- Proanthocyanidins
- Tannins
- Betalains
- Quinones and Xanthones
- Carotenoids
- Miscellaneous natural pigments

Characteristics of Natural Pigments

Pigment group	Number of compounds identified	Colour	Solubility
Anthocyanins	150	Orange, red blue	Water soluble
Flavonoids	800	Colourless, yellow	Water soluble
Proanthocyanidins	20	Colourless	Water soluble
Tannins	20	Colourless, yellow	Water soluble
Betalains	70	Yellow, red	Water soluble
Quinones	200	Yellow, black	Water soluble
Xanthones	20	Yellow	Water soluble
Carotenoids	450	Colourless, yellow, red	Lipid soluble
Chlorophylls	25	Green, brown	Organic solvents
Heme pigments	6	Red, brown	Water soluble
Riboflavin	1	Greenish yellow	Water soluble

Some Toxic Constituents of Plant Foodstuffs

Toxins	Major toxicity symptoms
• Protease inhibitors	Impaired growth and food utilization, Pancreatic hypertrophy
• Hemagglutinins	Impaired growth and food utilization, Agglutination of erythrocytes in vitro, Mitogenic activity to cell cultures in vitro
• Saponins	Hemolysis of erythrocytes in vitro
• Glucosinolates	Hypothyroidism and thyroid enlargement
• Cyanogens	HCN poisoning
• Gossypol pigments	Liver damage, Hemorrhage edema
• Lathyrogens	Osteolathyrism (skeletal deformities), Neurolathyrism (CNS damage)

Contd...

Contd...

Toxins	Major toxicity symptoms
• Allergens	Allergic responses in sensitive individuals
• Cycasin	Cancer of liver and other organs
• Favism	Acute hemonary anemia
• Phytoalexins	Pulmonary edema, Liver and kidney damage, Skin photosensitivity, Cell lysis in vitro
• Pyrrolizidine alkaloids	Liver and lung damage carcinogens
• Safrole	Carcinogens
• α-Amanitin	Salivation, Vomiting, Convulsions, Death
• Atractyloside	Depletion of glycogen

VISCOSITY

The consistency of a fluid is the property, which governs its flow characteristics. It is actually a measure of internal resistance to flow. At the present time, most of the devices used to measure the consistency of food materials show results that are expressed in terms of viscosity.

- Those fluids, which maintain a constant consistency irrespective of velocity, are known as Newtonian.

- The equation expressing the shear stress-rate of shear relationship of Newtonian fluids is given as $T = (-dV/dR) = \mu r$

  Where: T = Shear stress, $-dV/dR = r$ = rate of shear, μ = viscosity coefficient or viscosity.

- The classical expressions for non-time dependent non-Newtonian materials are as follows:

- **Bingham plastic:**

  $T = b (-dV/dR) + C$

  Where: b = proportionality factor, C = yield stress.

- **Pseudoplastic or power law:**

  $$T = b (-dV/dR)^s, \qquad O < S < 1$$

  Where: S = Pseudoplasticity constant, A pseudoplastic fluid offers less resistance to flow with increase in flow rate. Example of pseudoplastic fluid is solutions of gums.

- **Dilatant:**

  $$T = b (-dV/dR)^s, \qquad 1 < S < \infty$$

A dilatant fluid is one whose resistance to flow increases with flow rate or shear rate. Example of dilatant fluid is starch in water.

A general equation for all classical noontime dependent materials expressed as:

$$T = b \, (-dV/dR)^s + C$$

An equation, which fits many suspensions, is

$$\sqrt{T} = b \, \sqrt{-dv/dr} + \sqrt{C}$$

This equation is known as the caisson equation.

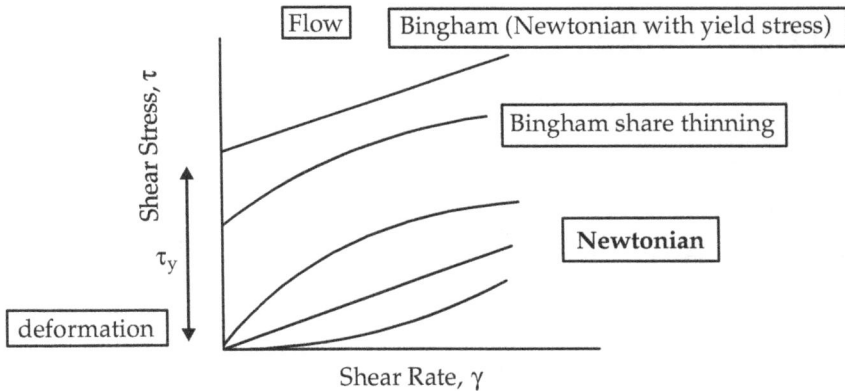

Fig. : Share stress-shear rate characteristics of various types of fluids.

- **Types of drying:**
  - Tunnel drying
  - Tray and compartment drying
  - Spray drying
  - Foam spray drying
  - Grain drying
  - Drum drying
  - Foam-mat-drying
  - Puff drying
  - Vacuum shelf drying
  - Sun drying
  - Freeze-drying

FREEZE-DRYING SOLID MATERIALS

In the freeze-drying of solid materials the rate of dehydration may be governed either by the rate of vapour diffusion or the rate of heat transfer to the ice surface, whichever is limiting? When a pressure build-up results, the sublimation temperature of the ice rises, reducing the rate of heat transfer to the ice surface in the interior of the material. There are two processes occurring during the freeze-drying of food materials possessing a definite structure. One involves the transfer of water vapour from the ice surface through the dried layer and the other, heat transfer from the exterior to the ice surface.

FOOD DISPERSIONS

Food systems divided into two major categories intact edible tissue and food dispersions.

Diphase Food Dispersions

Phase	Name of dispersion	Example
Solid + Liquid	Sol	Skim milk
Liquid + Liquid	Emulsion	French dressing
Gas + Liquid	Foam	Meringue
Gas + Solid	Solid foam	Foam candy
Solid + Gas	Solid aerosol	Smoke for flavouring food

As the temperature increases the surface tension decreases. As the concentration of the solution increases the surface tension (dynes/cm) also increases. The lowest surfactant concentration at which the micelles form is known as the "critical micelle concentration". It falls between 0.004 and 0.15 moles/liter. The concentration of monoglyceride increases the inner facial tension decreases.

FOAMS

Foam is a dispersion of gas bubbles in a liquid or semisolid phase. The typical food foams include whipped cream, ice cream, cake, bread, marshmallow, meringue and the forth-on beer. Presence of a foaming agent in the continuous phase must be prior to dispersion of gas. Foaming agents are surface-active liquids, glucosides, cellulose derivatives and proteins. Increasing the elasticity of the bubble wall by increasing the viscosity of the solution can enhance foam stability. For destruction of foam antifoam agent added in some food products for removing the foam at the level of 10 - 100 ppm e.g. dimethyl polysiloxanes (silicone oils).

AMINO ACIDS

Tryptophan is destroyed in the acid hydrolysis. The tryptophan is interacting with aldehyde like materials and resulting products are not observed on the amino acid analyzer.

Cysteine and cystine generally oxidize with performic acid and then do the HCl hydrolysis to get both cys and cys-cys as cysteic acid. One mole of cysteic acid per mole of cysteine and two moles of cysteic acid per mole of cystine.

Methionine and methionine sulfone may appear on the amino acid profile; with performic acid-oxdized samples, the conversion to methionine sulfone is essentially quantitative.

The spectrophotometer gives reading at 440 nm for proline and hydroxyproline and at 570 nm to other amino acids with ninhydrin reaction.

Precipitating the proteins with a compound, trichloroacetic acid (TCA) or perchloric acid (PCA) are two the most common. Kjaldhal then measures the soluble supernatant for its total nitrogen content, which is the non-protein nitrogen. In the soluble phase the free amino acids and that is non-protein nitrogen and also di and tripeptides are almost certainly in the soluble phase.

Major Complex Biomolecules of the Cells

Biomolecule	Building block (repeating unit)	Major functions
Protein	Amino acids	Fundamental basis of structure and function of cell (static and dynamic functions)
Deoxyribonucleic acid (DNA)	Deoxyribonucleotides	Repository of hereditary information
Ribonucleic acid (RNA)	Ribonucleotides	Essentially required for protein biosynthesis
Polysaccharide (glycogen)	Monosaccharides (glucose)	Storage form of energy to meet short term demands
Lipids	Fatty acids, glycerol	Storage form of energy to meet long term demands; structural components of membranes

Chemical Composition of a Normal Man

Constituent	Percent (%)	Weight (Kg)
Water	61.6	40
Protein	17.0	11
Lipid	13.8	9
Carbohydrate	1.5	1
Minerals	6.1	4
Total	100	66

Elements Composition of Proteins

Element	Percent content
Carbon	50 - 55
Hydrogen	6 - 7.3
Oxygen	19 - 24
Nitrogen	13 - 19
Sulphur	0 - 4
It also contain other elements in minor quantity such as	P, Fe, Cu, I, Mg, Mn, Zn, etc.

Colour Reactions Occur with Amino Acids/Proteins

Reaction	Specific group or amino acid
Biuret reaction	Two peptide linkages
Ninhydrin reaction	α-Amino acids
Xanthoproteic reaction	Benzes ring of aromatic amino acids (Phe, Tyr, Trp.)
Milons reaction	Phenolic group (Try)
Hopkins-Cole reaction	Indole ring (Trp)
Sakaguchi reaction	Guanidino group (Arg)
Nitroprusside reaction	Sulfhydryl group (Cys)
Sulfur test	Sulfhydryl group (Cys)
Pauly's test	Imidazole ring (His)
Folin-Coicalteau's test	Phenolic groups (Tyr)

Major Organs of Gastrointestinal Tract With Their Major Functions in Digestion and Absorption

Organ	Major function(s)
Mouth	Production of saliva containing α-amylase; partial digestion of polysaccharides
Stomach	Elaboration of gastric juice with HCl and proteases; partial digestion of proteins
Pancreas	Release of $NaHCO_3$ and many enzymes required for intestinal digestion
Liver	Synthesis of bile acids
Gall bladder	Storage of bile acids
Small intestine	Final digestion of foodstuffs; absorption of digested products
Large intestine	Mostly absorption of electrolytes; bacterial utilization of certain non-digested and/or unabsorbed foods

Absorption Rate of Sugars in Human Body

Monosaccharide	*Absorption rate (in duodenum and upper jejunum)*
Glucose	100
Galactose	110
Fructose	43
Mannose	20
Xylose	15
Arabinose	9

Blood Coagulation Factors Presents in Human Body

Factor	*Subunit molecular weight (Dalton)*
Fibrinogen	340,000
Prothrombin	720,000
Tissue factor, thromboplastin	370,000
Calcium (Ca^{++})	-
Proaccelerin, labile factor	330,000
Proconvertin, serum prothrombin conversion accelerator (SPCA)	50,000
Antithemophilic factor A, antithemophilic globulin (AHG)	330,000
Christmas factor, antihemophilic factor B	56,000
Plasma thromboplastin component (PTC)	-
Staurt-Prower factor	56,000
Plasma thromboplastin antecedent (PTA)	160,000
Hageman factor	80,000
Fibrin-stabilizing factor (FSF), fibrinoligase, Liki Lorand factor	320,000
Prekallikrein	88,000
High molecular weight kininogen (HMK)	150,000

pH of Important Biological Fluids

Fluid	*pH*
Pancreatic juice	7.5 - 8.0
Blood plasma (or whole blood)	7.35 - 7.45
Cerebrospinal fluid	7.2 - 7.4
Tears	7.2 - 7.4
Intestinal fluid	7.2 - 7.4
Human milk	7.2 - 7.4
Saliva	6.4 - 7.0
Intracellular fluid (cytosol)	6.5 - 6.9
Gastric juice	1.5 - 3.0
Urine	5.0 - 7.5

Energy Reserves of a Normal Man (70 kg)

Energy source (main storage tissue)	Weight (kg)	Energy equivalent (Cal)
Triacylglycerol (adipose tissue)	15	135,000
Protein (muscle)	6	24,000
Glycogen (muscle, liver)	0.2	800

Energy Relationship in Major Mammalian Organs

Organ/tissue	Energy compound preferably utilized	Energy compound exported
Liver	Amino acids, glucose, fatty acids	Glucose, fatty acids, ketone bodies
Adipose tissue	Fatty acids	Fatty acids, glycerol
Skeletal muscle	Fatty acids, glucose	None, Lactate
Brain	Glucose, ketone bodies (in starvation)	None

Water Distribution in Adult Man (70 kg)

Component	% Body weight	Volume (L)
Total	60	42
Intracellular fluid (ICF)	40	28
Extra cellular fluid (ECF)	20	14
Interstitial fluid	15	10.5
Plasma	5	3.5

Electrolytes Present in Human Body Fluid (expressed as mEq/L)

Extra cellular fluid (Plasma)				Intracellular fluid (Muscle)			
Cations		Anions		Cations		Anions	
Na^+	142	Cl^-	103	K^+	150	HPO_4^-	140
K^+	5	HCO_3^-	27	Na^+	10	HCO_3^-	10
Ca^{++}	5	HPO_4^-	2	Mg^{++}	40	Cl^-	2
Mg^{++}	3	SO_4^-	1	Ca^{++}	2	SO_4^-	5
		Proteins	16			Proteins	40
		Organic acids	6			Organic acids	5
	155		155		202		202

Enzymes Take Part in the Diagnosis of Various Types of Diseases

Serum enzyme (elevated)	Disease (most important)
Amylase	Acute pancreatitis
Serum glutamate pyruvate transaminase (SGPT)	Liver diseases (hepatitis)
Serum glutamate oxaloacetate transaminase (SGOT)	Heart attacks (myocardial infarction)
Alkaline phosphatase	Rickets, obstructive jaundice
Acid phosphatase	Cancer of prostate gland
Lactate dehydrogenase (LDH)	Heat attacks, liver diseases
Creatine phosphokinase (CPK)	Myocardial infarction (early marker)
Aldolase	Muscular dystrophy
5'-Nucleotidase	Hepatitis
γ-Glutamyl transpeptidase (GGT)	Alcoholism

Different Types of Diabetes Mellitus

Character	Insulin-dependent diabetes mellitus (IDDM)	Non-insulin dependent diabetes mellitus (NIDDM)
General		
• Prevalence	10-20% of diabetic population	80-90% of diabetic population
• Age at onset	Usually childhood (< 20 yrs)	Predominantly in adults (< 30 yrs)
• Body weight	Normal or low	Obese
• Genetic predisposition	Mild or moderate	Very strong
Biochemical		
• Defect	Insulin deficiency due to destruction of β-cells	Impaired in the production of insulin by β-cells and/or resistance of target cells to insulin
• Plasma insulin	Decreased or absent	Normal or increased
• Auto antibodies	Frequently found	Rare
• Ketosis	Very common	Rare
• Acute complications	Ketoacidosis	Hyperosmolar coma
Clinical		
• Duration of symptoms	Weeks	Months to years
• Diabetic complications at diagnosis	Rare	Found in 10-20% cases

Contd...

• Oral hypoglycemic drugs	Not useful for treatment	Suitable for treatment
• Administration of insulin	Always required	Usually not necessary

DIETARY GUIDELINES FOR PREVENTION OF DISEASES

- There are many common features among diseases insofar as the influence of dietary factors on disease prevention is concerned.

- Based on these observations an integrated or common dietary guideline for prevention of these groups of diseases (obesity, diabetes, coronary heart diseases, cancer) and for maintaining health can be provided:

1. Excess calorie intake should be avoided and energy balance and constant body weight should be maintained. Excess energy intake can be prevented in several ways:

   (a) Reducing total dietary intake

   (b) Increasing energy expenditure through exercise

   (c) Reducing total fat content in the diet and

   (d) Increasing dietary fiber content

   Maintaining of energy balance is essential for prevention of most of these diseases.

2. Protein intake can be maintained to provide 10 to 12 percent of energy.

3. Energy contribution from carbohydrate should be around 60 percent and most of it should be in the form of complex carbohydrate (starch) and contribution from simple sugars should be less than 10 percent of energy.

4. The amount of fat in the diet should contribute less than 30% energy and preferably around 25% cholesterol content should be reduced to the minimum.

5. Vitamin and mineral intake should be adequate to meet the normal requirement according to recommended dietary allowances and particular attention should be paid to adequate intake of Ca, Mg, Zn and Selenium and vanadium. Sodium intake should be kept low (2 to 3 g/day) and potassium intake should be increased.

6. Diet should contain an adequate level of fiber covering all types of fibers of fibers in order to obtain an optional response.

Food Adulterants and Simple Tests to Detect Food Adulteration

Substance	Adulterant	Tests
Turdal	Lakh dal or metanil	• Lakh dal is irregular in shape and of lighter colour than turdal. • Add concentrated HCl to moisten dal yellow colour will turn into magenta red if metanil yellow is present.
Rawa	Iron fillings to add	• Pass magnet through the rawa. Iron fillings will weight cling to it.
Sago	Sand and talcus	• Pure sago swells and on burning, it leaves hardly any ash.
Bajra	Fungus	• Immerse in salt water, fungi will come on top.
Jaggery	Metanil yellow	• HCl added to jaggery solution turns magenta coloured.
Bura sugar	Washing soda	• Gives effervescence with HCl if washing soda is present. • If dissolved in H_2O washing soda will turn red litmus blue.
Honey	Commercial invert sugar	Fiehe's test. Mix about 5 g of honey with 10 ml of ether in a mortar, using a pestle. Decant off the ether extract into a china dish. Repeat more 3 times and collect the extract together. Evaporate the solution and add 15 solution of freshly sublimed resorcinol in concentrated hydrochloric acid. Immediately appearance of a cherry-red colour indicates presence of invert sugar.
Ghee or butter	Vanaspati	• Dissolve 1 teaspoon of sugar in 10 ml of HCl. Add 10 ml of melted ghee and shake thoroughly for 1 min. Allow it to stand for 10 min. If vanaspati has been added the aqueous layer will be red.
Coconut oil	Any other oil	• Place a small quantity of oil in refrigerated coconut oil will solidify leaving the adulterant as a separate layer.
Edible oil	Argemone oil	• On treatment with nitric acid it will give red colour in acid layer, indicating the presence of argemone oil.
Milk	Water	Measure specific gravity with lactometer, normal reading 1.1030 to 1.1034.
Milk or Curd	Cane sugar	Add 0.1 g of resorcin and 1 ml of concentrated hydrochloric acid to 10 ml of the sample and boil. A rose red colour indicates addition of sugar.

Substance	Adulterant	Tests
Milk, Curd, the Khoa, Ghee, Butter	Starch	Add a drop of iodine solution to a little of sample. Blue colour indicates added starch in any form.
Tea powder	Exhausted tea leaves, dried powder and artificially coloured	• Sprinkle the powder on a wet white blotting paper. Spots of yellow and red colour appearing on paper indicated that tea is artificially coloured.
Coffee	Chicory	• Shake a small portion in cold water. Coffee will float while chicory sink retaining the water brownish red.
Coffee powder	Starch (toasted bread crumbs, rye, wheat peas etc.).	Make decoction of the coffee, decolorize it by adding potassium permanganate and then add a drop of iodine solution. Blue colour indicates starch.
Soft drink	Mineral acid other than phosphoric acid	Soak a strip of filter paper in a 0.1% solution of metanil yellow and then dry. Dip one end of paper into the soft drink. Wetted portion turns violet if mineral acid is present.
Cardamom	Oil is removed and pods are coated with talcum powder	• On rubbing talcum will stick to the fingers. On testing, if there is hardly any aromatic flavour, it indicates removal of essential oil.n
Black pepper	Papaya seeds	• Papaya seeds are shrunken, oval and grayish brown.
Asafoetida	Resin or gum (scented and coloured)	• Pure asafetida dissolves in water to form a milky white solution.
		• Pure asafetida burns with bright flame on being ignited.
Carraway seeds	Grass seeds	Grass seeds are smaller than caraway and they have no smell and taste.
Cinnamon	Wood bark	• It is far harder than cinnamon and may not have aroma and smell of cinnamon.
Cloves	Oil may be removed	• If so, cloves appear shrunken, Nagkesa will not give the taste of cloves.
Cumin seeds	May contain grass seeds coloured with charcoal dust	• If rubbed in hand, fingers will be black.
Mustard seeds	Argemone seeds	• Argemone seeds have no round structure; they are pointed and are more blackish than mustard seeds.
Chilli powder	Saw dust & red colour	• Sprinkle on the surface of water saw dust floats. Added colour will colour the water.
Saffron	Maize fibers dried coloured & scented	• Genuine saffron is tough will not break easily like that of artificial saffron.

Substance	Adulterant	Tests
		• Saffron dissolves easily in water giving aroma.
Turmeric	• Metanil yellow • Starch	• When conc. HCl is added to solution of turmeric powder, it turns magenta if metanil yellow is present.
		• Add iodine solution to turmeric solution it will turn violet if starch is present.
Coriander	Horse dung powder	• Soak in water, horse dung will float which can be easily detected.
Betel nuts	Sawdust	• Sprinkle in water the wood shavings will float and the added colour will come off in water.
Pan masala	Saccharin	• Saccharin is bitter in taste.

F. P. O. Specifications For Fruits and Vegetable Products

Product	Specifications	
	Minimum % of total soluble solids in final product (w/w)	*Minimum 5 of fruit juice/prepared fruit in final product (w/w)*
Fruit syrup	65	25
Crush	55	25
Squash	40	25
Fruit nectar (excluding orange and pineapple nectars)	15	20
Orange/pineapple nectars	15	40
Mango nectar	15	20
Cordial	30	25
Unsweetened juice	Natural	100
Sweetened juice	10	85
Ready-to-serve fruit beverage including aerated water containing fruit juice	10	10
Barley water	30	25 (barley starch 0.25%)
Fruit juice concentrate	32	100
Jam and fruit cheese	68	45
Fruit jelly and marmalade	65	45
Fruit preserve	68	55
Fruit chutney	50	40
Synthetic syrup/sarbat (from herbs, flowers, essences)	65	-
Candied and crystallized or glazed fruit and peel	Total sugar % not less than 70%	Reducing sugars % of total sugars not less than 25%

WHAT ARE THE UNIT OPERATIONS AFTER HARVESTING OF FOOD GRAINS?

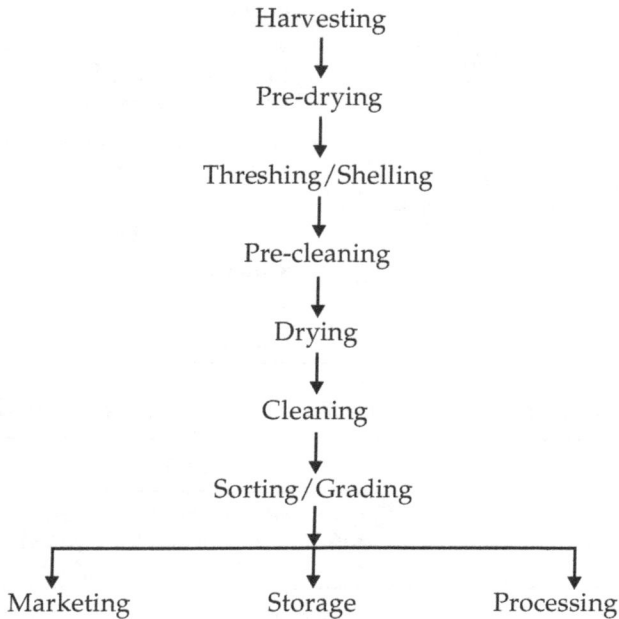

Harvesting
↓
Pre-drying
↓
Threshing/Shelling
↓
Pre-cleaning
↓
Drying
↓
Cleaning
↓
Sorting/Grading
↓

Marketing   Storage   Processing

HOW CEREAL GRAINS, MILLETS AND PULSES CAN BE PROCESSED?

- Cleaning
- Dehulling/decortication
- Malting
- Fermentation
- Processing of cereal grains, millets and pulses with various solvents
- Flour preparation
- Semolina/Rava/Grits preparation
- Fortification of cereals, millets and pulses
- Value-added health food products preparation
- Pulses and oilseeds processing items:
    - Full fat and defatted oilseed flours
    - Composite flours; protein concentrates and isolates; weaning foods

- Cereal-oilseed-pulses flour based traditional foods
- Oil extraction
- Refining of oil

What is the various food products preparation can be done from cereals, millets and pulses?

WHEAT

1. Traditional products
2. Bakery products
3. **Industrial products :** Like whole-wheat flour, refined flour, semolina, wheat bran, amylase rich foods, macaroni products, wheat malt extracts etc.

RICE

1. Traditional products

   Industrial products: Like rice hulls/husk, rice bran, rice bran oil, puffed rice, flaked rice, quick cooked rice, parboiled rice etc.

CORN / MAIZE

1. Traditional products
2. **Commercial products:** Like corn grits, corn flakes, corn flour, pops, corn starch, breakfast cereals etc.

SORGHUM

1. Ethnic products
2. Bakery products
3. Value-added products
4. Industrial products: Starch, fiber, semolina etc.

PEARL MILLET

1. Traditional products (flour, bhakari)
2. Nutraceutical/Health products
3. Industrial products : Fibers, mineral source etc.

FINGER MILLET

1. Traditional products
2. Health products
3. Industrial products

FOXTAIL MILLET

1. Traditional products
2. Health products
3. Industrial products

Prosomillet, Little millet, Barnyard millet, and Kodo millet : 1. Traditional/ Ethnic products, 2. Nutraceutical/Health products, 3. Industrial products etc.

PULSES

1. Dehulling
2. Dhal preparation
3. Flour preparation (Besan preparation)
4. Concentrates and isolates

OILSEEDS

1. Oil
2. Defatted meal
3. Full fat meal
4. Protein concentrates
5. Protein isolates
6. Cakes
7. Weaning foods
8. Oil refining
9. Composite flours
10. Cereal-pulses-oilseed based traditional foods

SUGARCANE

1. Sugar
2. Baggase

3. Molasses
4. Alcohol
5. Jaggery
6. Syrup

COTTON

1. Thread
2. Cloths
3. Oil
4. Cakes
5. Fuel

PROCESSING OF FRUITS AND VEGETABLES INTO VARIOUS PRODUCTS

- By drying, dehydration and concentration.

  Sun drying, solar drying, shade drying, osmotic dehydration etc.

- Freezing of fruits and vegetables

1. Sharp freezing (slow freezing)
2. Quick freezing

   (a) by direct immersion

   (b) by indirect contact with refrigerant

   (c) by air blast

   1. Fluidized bed freezing

   2. Plate freezing

   3. Cryogenic freezing

   4. Dehydro-freezing

   5. Freeze-drying

- Processing by fermentation:

  Fermented beverages: Wine, champaigne, port, sherry, tokay, Muscat, perry, orange wine, berry wine, nira, feni, cider etc.

- Vinegar:

  Brewed vinegar: Fruit vinegar, potato vinegar, malt vinegar, molasses vinegar, honey vinegar, spirit vinegar, spiced vinegar etc.

  Artificial vinegar: 4% acetic acid vinegar.

PROCESSING OF FRUITS AND VEGETABLES INTO UNFERMENTED BEVERAGES

- Juice
- Ready-to-serve (TRS)
- Nectar
- Cordially, Squash
- Crush
- Syrup
- Fruit juice concentrate
- Fruit juice powder
- Barley water
- Carbonated beverages

PROCESSING OF FRUITS AND VEGETABLES INTO VARIOUS FOOD PRODUCTS

- Jam
- Jelly
- Marmalade
- Preserve
- Candied
- Crystallized fruits and vegetables
- Pickles: With salt, vinegar, oil and mixture
- Chutneys and sauces/ketchups
- Tomato processing
  - Juice, puree, paste, sauce/ketchup, chutney, cocktail, soup, pickles etc.
- Potato processing:
  - Chips/wafers
  - French fries
  - Flour
  - Canning

- Mushroom processing
    1. Canning
    2. Dehydration
        (a) Freeze-drying
        (b) Freezing
    3. Steeping preservation

VALUABLE PRODUCTS FROM FRUITS AND VEGETABLES

1. Sauerkrant
2. Mango slices (Amchur)
3. Mango leather
4. Fruit cheese
5. Fruit butter
6. Fruit toffee
7. Fruit candy
8. Papain
9. Pectin
10. Fruit pulp powder

PROCESSING OF BEVERAGES

It includes coffee, tea, cocoa, non-alcoholic beverages, alcoholic beverages and beverages based on fruit juices.

- **Coffee**
    - Roasting
    - Vacuum coffee
    - Drip coffee
    - Percolator coffee
    - Steeped coffee
    - Espresso coffee
    - Iced coffee
    - Soluble coffee

- **Tea**
  - Black tea
  - Green tea
  - Oolong tea
  - Iced tea
  - Instant tea
- **Cocoa**
  - Chocolate
  - Cocoa beverage
  - Cocoa butter
- **Alcoholic beverages**
  - Beer:
    - Bitter
    - Mild
  - Cidey:
    - Dry
    - Sweet
  - Wine:
    - White
    - Red
    - Fortified
  - Spirit:
    - Whisky
  - Liquor:
    - Benedictine

PROCESSING OF SUGARCANE AND SUGAR BEET

- Sugar from sugarcane and sugar beet:
  1. Raw sugar
  2. Refined sugar
  3. White sugar
  4. Cube sugar

5. Powdered sugar
6. Brown sugar
7. Rock sugar
8. Diamond sugar

- Jaggery from sugarcane
- Molasses
- Alcohol
- Syrups:
  1. Cane syrup
  2. Corn syrup
  3. High fructose syrup
  4. Maple syrup
  5. Sweet sorghum syrup
- Honey:
  1. Fructose about 38%
  2. Glucose 31%
  3. Sucrose 2%

PROCESSING OF MILK INTO VARIOUS PRODUCTS

- Curd (Dahi)
- Buttermilk
- Butter
- Ghee
  1. Evaporated milk
  2. Condensed milk
  3. Toned milk
  4. Dry milk
- Cheese
  1. Cottage cheese
  2. Cream cheese
  3. Other cheese (Swiss cheese)
  4. Panir (paneer)

5. Decca cheese

6. Surti cheese

7. Bandal cheese

8. Channa

- Traditional milk preparations

    1. Khoa

    2. Gulab jamun

    3. Basundi

    4. Ice-cream

    5. Srikhand

- Milk beverages

    1. Skim and low-fat milks

    2. Concentrated milks

- Homogenized milk

    1. Canned and frozen whole milks

    2. Soft-cured milk

    3. Low sodium milk

    4. Malted milk

    5. Cultured milk

    6. Flavoured milk and milk drinks

- Filled and imitation milks

- Certified milk

- Fermented milk

    1. Curd

    2. Yoghurt

    3. Kumiss

    4. Leben

    5. Kefir

    6. Matzoon

    7. Gioddin

8. Taette
9. Skyr

PROCESSING OF MEAT

- Canned meat
- Cured meat
- Smoked meat
- Sausages
- Dehydrated meat
- Strained baby foods based on meat

PROCESSING OF POULTRY

- Chilling of slushed dice
- Freezing
- Canning
- Sausages
- Dehydration
- Canning of fried chicken
- Cured and smoked poultry

PROCESSING AND PRESERVATION OF FISH

- Freezing
- Canning
- Fish meal (animal and poultry feed)
  1. Wet process (fatty fish)
  2. Dry process (lean fish)
- Fish protein concentrate
- Fish liver oil

  Preservation methods for meat and poultry:
- Refrigeration
  1. Chilling

- Freezing
- Thermal processing cooking
  1. Canning
- Drying
  1. Dehydration
- Freeze-drying
- Curing and smoking
- Chemical preservation
  1. Antibiotics
  2. Chlorine
  3. Nitrates/Nitrites
  4. Polyphosphates
  5. Edible organic acids
     (a) Sorbic acid
     (b) Succinic acid
     (c) Lactic acids
- Ionizing radiation (Beta or gamma)
- Pickles

PROCESSING OF EGGS

- Liquid whole eggs
- Liquid yolks
- Liquid albumin
- Frozen eggs
- Egg solids
- Egg powder
  1. Spray method
  2. Pan method
- By products:
  1. Albumin flakes
  2. Eggshell meal

PRESERVATION OF SHELL EGGS (STORAGE)

- Wet immersion methods
    1. Limewater or lime sealing method (Ca $(OH)_2$ + CO_2 → $CaCO_3$ + H_2O)
    2. Water glass method (10% Sodium silicate)
- Dry method:
    1. Oiling
    2. Dip method
    3. Spraying method
    4. Gaseous atmosphere
- Thermo stabilization or heat treatment methods:
- Cold storage or refrigeration:
    1. Long-term storage (-1.1°C, RH 85-90%)
    2. Short-term storage (-12.8°C, RH 60-70%)

WHAT ARE THE BY-PRODUCTS OF OILSEEDS?

- Soybean meal
- Groundnut meal
- Cotton meal
- Sesame seed meal
- Coconut meal
- Sunflower seed meal
- Rapeseed meal
- Protein isolates from oilseeds and nuts
- Food analogue
    1. Textured vegetable proteins: Spun vegetable protein
    2. Meat and dairy analogues from vegetable proteins
- Fermented soybean products
- Irradiated and radiated foods

What are the marketing/procurement channels for food grains?

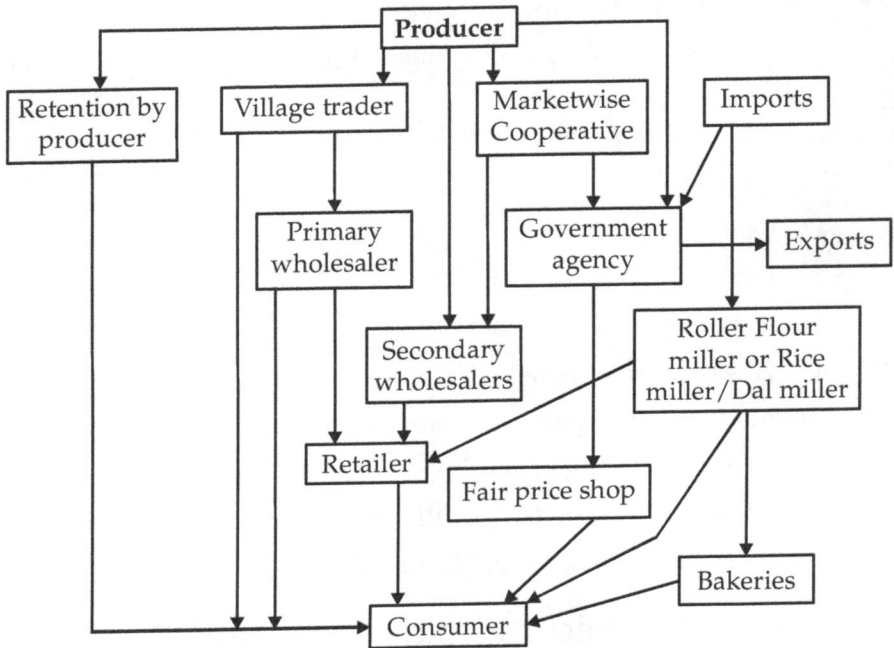

What are the marketing institutions present in India (Procurement, distribution institutions)?

A. Public sector institutions

1. Directorate of Marketing and Inspection (DMI)
2. Commission for Agricultural Costs and Prices (CACP)
3. Food Corporation of India (FCI)
4. Cotton Corporation of India (CCI)
5. Jute Corporation of India (JCI)
6. Specialized commodity Boards

   - Coffee Board
   - Coconut Board
   - Cardamom Board
   - Coir Board
   - Tobacco Board

   - Rubber Board
   - Tea Board
   - Spices Board
   - Oilseeds and Vegetable Oils Board
   - Silk Board

- Arecanut Board
- National Horticulture Board (NHB)
- National Dairy Development Board (NDDB)

7. Others

- Central Ware Housing Corporation (CWC)
- State Ware Housing Corporation (SWCS)
- State Trading Corporation (STC)
- Agricultural and Processed Food Export Development Authority (APEDA)
- Export Inspection Council (EIC)
- Marine products Export Development Authority (MPEDA)
- Silk-export Promotion Council (SEPC)
- The Cashew Nuts Export Promotion Council of India (CEPCI)
- Agricultural Produce Market Committees (APMC)
- State Agricultural Marketing Boards (SAMB)
- Council of State Agricultural Marketing Boards (COSAMB)
- State Directorates of Agricultural Marketing (SDAM)
- Research Institutions and Agricultural Universities (RIAU)

B. CO-OPERATIVE SECTOR INSTITUTIONS

- National Cooperative Development Corporation (NCDC)
- National Agricultural Cooperative Marketing Federation (NAFED)
- National Cooperative Tobacco Growers' Federation (NTGF)
- National Consumers Cooperative Federation (NCCF)
- Tribal Cooperative Marketing Federation (TRIFED)
- State Cooperative Marketing Federations (SCMF)
- Primary Agricultural Cooperative Marketing Societies (PACMS)
- Special Commodity Cooperative Marketing Organizations (sugarcane, cotton, milk, grape, pomegranate, guava etc.) (SCCMO).

Utilization/Processing of Fruits and Vegetable Waste

Fruit/Vegetable	Waste	Utilization/Processing into
Apple	Pomace, cores	Pectin, cider, vinegar, chutney
Apricot	Kernels	Jam, oil, pharmaceutical and cosmetic preparations
Citrus fruits	Peels, rags, seeds, sludge	Candy, oil, pectin, marmalade, toffee, cattle feed, vinegar, citric acid
Grape	Stems, pomace, seeds	Cream of tartar, oil, cattle feed, jelly, chutney
Guava	Cores, seeds, peels	Cheese
Jack fruit	Rind, seeds	Jelly, pectin, seed flour
Mango	Peel, kernels	Vinegar, kernels powder
Passion fruit	Rind, seeds	Pectin, oil
Peach	Kernels	Oil for industrial use
Pear	Peels, cores	Animal feed, perry, vinegar
Pineapple	Shell, trimmings, cores, tops, leaves	Juice, citric acid, alcohol, vinegar, candy, cattle feed etc.
Other fruits	Banana peel, pseudostem, latex	Banana cheese, paper pulp, papain, proteolytic enzyme etc.
Tomato	Seeds, trimmings	Oil, juice, sauce, puree
Other vegetables	Waste potato, cabbage, cauliflower, turnip, carrot, beet, beans, sweet potato etc.	Cattle feed.

6

FOOD BIOTECHNOLOGY

Improved nutritional properties mostly quality and safety by using biotechnology of cereals: Cereals are very important part of human diets. The three major species, wheat, maize and rice for a large proportion of the calories and protein in human diets. The importance of cereals in the food chain is also attribute to the extensive use of cereals in the diets of animals. The major constituents of cereals are the carbohydrates and proteins. Other grain components such as lipids and vitamins may be of great significance in human nutrition because of the large contribution of cereals to the diet. Biotechnology provides new options for manipulation of the nutritional properties of cereal grains. The carbohydrates of cereals include the simple sugars, the more complex oligosaccharides such as fructans, storage polysaccharides of the grain (starch) and cell wall polysaccharides, all of which are of nutritional value. All of these carbohydrate components are potential targets for manipulation in improvement of cereal quality. For example, sugar beet has been transformed to produce fructans. Benefits that may result include reduced cariogenic bacterial (dental health), lower energy value and stimulation of beneficial bacteria in the colon. The sugar content may also influence the quality of the grain for various products. Fructans may be considered to be important to human nutrition because of their possible role as soluble fiber. Starch as the major component by weight of the grain may have a great impact on nutritional quality. Resistant starches (not digested in the gut) may be considered critical in influencing the incidence of certain human diseases, such as heart disease. The cell wall polysaccharides may also be important as either soluble or insoluble fiber, depending on the composition of the polysaccharides in the cereal product. Soluble fibers may reduce the risk of heart disease while insoluble fibers contribute to reduce risk of colonic cancers. Cereal proteins are not well balanced in amino acids required in a nutritionally balanced diet, and genetic engineering may provide opportunity to improve the balance of essential amino acids in cereal-based diets. The lipids in cereals are generally of limited importance in human nutrition but may be important in animal diets. The manipulation of iron levels in cereals through the introduction of haemoglobin illustrates the potential application of biotechnology to enhancing the nutritional value of cereals.

What are the future prospects and limitations of biotechnology in cereals?

Biotechnology is likely to have a major impact on the value of cereal production both by increasing productivity and by improvements in product quality. The improved productivity is likely to result initially from the removal of biotic stress constraints associated with major pests and diseases. Herbicide resistance is a potion that is likely to be able to achieve early adoption and success. Improvements in grain quality are likely to be generally more difficult to achieve. The major attraction of biotechnology is the possibility of introducing totally new or novel characteristics into cereals that will result in products with characteristics outside the range of those currently available. A major limitation to the introduction of such characteristics is the requirement of cereals to be compatible with existing processes of cereal food production. Market resistance to products requiring new processing techniques will come from the large investment that may be required to develop new processing facilities. Improved methods of quality control and analysis of product identity and purity will enhance the value of cereal products. Adverse consumer attitudes are also a significant risk if transgenic products are not well designed marketed.

Malting is a one of the most important field for the development using biotechnology stool. During malting normally about 20% of the starch sugar remains as dextrins. During malting proteases and carboxypeptidases operates to breakdown proteins and carbohydrates. Significant proteolysis is necessary to generate amino acids for yeast nutrition and to remove the bulk of proteins, which may later contribute to beer haze, but sufficient polypeptide needs to be consumed in order to allow foam formation when beer is served. In fermentation of the boiled, clarified ort, sugars, amino acids, and other nutrients are metabolized by brewers yeast Saccharomyces cerevisiae to yield new yeast mass, with ethanol and carbon dioxide as major end products. Proanthocyanidins are mostly responsible for haze formation in beer. Beer stored in oak barrels adds colour and leads to both the release of compounds positive to flavour such as furfural and tannins and the absorption of undesirable (primarily suplhur compounds).

What are the targets for genetic modifications of cereals for malting, brewing and distilling?

Following are the various target sections for genetically modifications.

S.No.	Target	Benefits
1.	Improved disease and pest resistance	Reduces chemical resistances in the grain
2.	Safe barley storage without drying	Energy savings
3.	Inhibitory to grain microflora	Improved malting and brewing properties and reduced tendency for gushing
4.	Reduced husk content	Higher yields
5.	Control of dormancy	Improved efficiency of malting operations
6.	Quicker and more even water uptake	Faster malting

Contd...

Contd...

S.No.	Target	Benefits
7.	Increased natural antioxidants, e.g., catechin and ferulic acid in the grain	Improved beer flavour stability
8.	Reduced levels of β-glucan in barley	Faster malting and easier wort separation and beer filtration
9.	Thermostable β-glucanase	More effective β-glucan breakdown in mashing
10.	Themostable α-amylase, β-amylase and limit dextrinase	Improved extract and fermentability
11.	Modified barley endosperm structure, e.g., changed amylose/amylopectin ratio or increased ratio of large to small starch granules.	Starch more easily convertible giving increased yields.
12.	Increased levels of hydrophobic proteins	Improved beer foam
13.	Reduced levels of heat-stable proteins that limit dextrinase	Improved fermentability
14.	Proanthocyanidin-free barley	Better haze stability
15.	Blockage of mthylation of methionine in germination	Lower dimethyl sulphide in larger.
16.	Stimulation of methylation of methionine in germination	Higher dimethyl sulphide in larger
17.	Lower levels of malt lipoxygenase	Better control of beer staling
18.	Sorghum more amenable to malting and brewing	Improved economy of brewing in Africa
19.	Cyanogenic glycoside-free grain	Elimination of carcinogenic ethyl carbamate production of distillation

What is mean by nutritional stress and which are the factors responsible for it?

The nutritional quality of foods depends on the total amount of nutrient present in the food consumed, as well as the levels of antinutritive factors.

Nutritional stress factors are defined as elements which increase human nutritional requirements by destroying essential nutrients, making essential nutrients unavailable, interfering with the utilization of essential nutrients, increasing with the digestion of food, and/or decreasing food intake.

Nutritional Stress Factors Presents in Various Types of Food

Stress factor	Chemical nature		Occurrence	Action	Dietary effect
Avidin	Protein	Egg white	Binds biotin	Biotin unavailable	
Goitrin	Glucosinolate	Rapeseed	Goiterogenic	Reduces iodine uptake	
Gossypol	Polyphenol	Cottonseed	Chelate metals	Anemia	

Contd...

Contd...

Stress factor	Chemical nature		Occurrence	Action	Dietary effect
Limarin	Cyanogenetic glucoside	Cassava	Releases cyanide	Potential poisoning	
Oxalates	Organic acid	Spinach, amaranth	Chelate cations	Calcium and iron unavailable	
Phytates	Organic acid	Cereals, legumes	Chelate metals	Reduce mineral availability	
Solanine	Alkaloid	Potato	Inhibits cholinesterase	Gastrointestinal or neurological disorders, potential poisoning	
Tannins	Polphenols	Beans	Bind proteins	Proteins insoluble, inactive enzymes. Reduce iron and B_{12} availability	
Thiaminases	Protein	Fish	Destroy thiamine	Possible thiamine deficiency	
Trypsin inhibitors	Protein	Legumes, cereals	Inhibit proteolysis	Reduce protein digestibility	

In which food industry biotechnologically modified proteolytic enzymes can be used?

Biotechnologically modified proteolytic enzymes can be used in following various food industries for various applications.

S.No.	Food Industry	Use of proteolytic enzyme
1.	Antinutrients	Remove phytate, gossypol, nucleic acids and enzyme inhibitors
2.	Baked foods	Softening action in doughs, reduce mixing time, increase extensibility, improve texture and grain loaf volume, liberate β-amylase
3.	Brewing	Body, flavour and nutrient development during fermentation and in filtration and clarification, chill proofing
4.	Cereals	Increase drying rate of proteins, improve product handling characteristics, production of miso and tofu
5.	Cheese	Coagulate casein, flavour development during aging
6.	Chocolate, cocoa	Facilitate fermentation
7.	Egg, egg products	Improve drying products
8.	Feeds	Waste product conversion to feed, digestive aid
9.	Fish	Solubilization of fish protein concentrates, recover oil or meat scraps from inedible parts
10.	Legumes	Hydrolyzed protein products, remove flavour, plastein formation
11.	Meats	Tenderization, recover proteins from bones
12.	Milk	Coagulate rennet puddings, prepare soybean milk
13.	Protein hydrolysates	Soy sauce, fish sauce, bouillon, dehydrated soups, gravy powders, processed meats, special diets.
14.	Wine	Clarification, decrease foaming, promote malolactic fermentation

What are the problems facing the cereal biotechnology industry?

- **List of problems:**
    1. Consumer acceptability.
    2. Selectable marker, Antibiotic resistances.
    3. Selectable markers to be removed from plants after the transformation process.
    4. Inadvertently introduced or produced toxins in transgenetic plants.
    5. Spread of transgenes through pollen to related wild plants.
    6. The potential negative effects of genetically modified crops on biodiversity.

WHAT IS FERMENTED FOODS?

Natural growth of certain microbes on grains, mostly cereals, improved their flavour, texture, and nutritional value. This promoted detailed studies on the various conditions required for obtaining quality products. After that, the microorganisms were isolated and their strains were improved for the fermentation technology. Most of the foods are produced by solid substrate fermentation and several of them are processed on an industrial scale.

What are the problems in solid substrate fermentation ?

They are difficult to scale up, it is difficult to monitor and control the various factors such as temperature, pH, nutrient distribution during the process, gas exchange such as O_2 supply and CO_2 removal, heat transfer and removal are not easy and efficient as with liquid fermenters.

What are the advantages of solid substrate fermentation?

1. Improved flavour, 2. Elimination of undesirable flavours, 3. Improvement in the texture of food, 4. Enhancement in nutritional value higher level of proteins, minerals, vitamins and even antibiotics, 5. Reduce cooking time, 6. Increased digestibility, 7. Some fermented foods are used as protein rich foods as well as meat supplement.

WHICH ARE THE ENZYMES USED IN FOOD INDUSTRY?

Food industry utilizes various types of enzymes for processing of various foods such as production of various types of syrups from starch or sucrose (α- and β-amylases, glucoamylase, pullulanase, invertase, glucose isomerase, proteases, removal of glucose and or molecular oxygen using glucose oxidase and catalase, use of lactase in dairy industry and use of enzymes in fruit juice and brewing industries).

Common Enzymes Used in Food Processing Industries

S.No. Enzyme	Substrate	Action/Objective
A Proteases		
1. Endogenous proteases	Meat	Tenderization of meat, flavour development
2. Subtilisin	Soya protein	Partial hydrolysis; increase whipping expansion, emulsifying capacity; hydrolysate may be added to cured meats.
3. Neutral or alkaline	proteases	Animal & fish bones (contain about 5% meat) Mashed bone incubated at 60°C for 4 h; meat slurry used in cannel meats and soups.
4. Subtilisin	Red blood cells	RBCs hameolysate subjected to hydrolysis; haeme molecules precipitate and are removed; purified hydrolysate spray dried and used in cured meats, sausages, luncheon meats etc.
5. Papain (in active form)	Meat tenderization	Injected into the jugular vein shortly before slaughter; after slaughter, papain is activated and tenderizes the meat; only 2-5 ppm (body weight) enzyme injected.
6. Heat labile fungal protease	Dough from high gluten wheat varieties	Hydrolysis of gluten; makes dough suitable for biscuit, pie pastry making.
B. Glucose oxidase	D-glucose	D-glucose oxidized to gluconic acid, O_2 utilized, H_2O_2 produced.
C. Catalase	H_2O_2	Degrades H_2O_2 into water and O_2; used in combination with glucose oxidase to remove glucose and/or O_2 from foods, drinks etc.
D. Amylases (α- and β-), glucoamylase, pullulanase, invertase, glucose isomerase (immobilized)	Starch, sucrose, D-glucose)	Production of glucose, maltose and high fructose syrups.
7. Rennin	Milk	Coagulation of milk solids & convert into cheese

Enzymes Used in Fruit Juice and Beverage Industry?

Various types of enzymes are used in fruit juices and brewing industries to achieve specific objectives.

The cloudiness of fruit juices and wines is mainly due to pectins, which may exhibit various degrees of methyl esterification and are usually associated with other plant polymers and even cell debris. Pecteolytic enzymes prepared from A. niger digest the pectins. These enzymes preparations are mixtures of the polygalacturonase, pectin esterase, pectin lyase, and hemicellulase. These enzymes have pH optimum between 4 and 5 as well as optimum temperature below 50°C. These are generally added directly in to the fruit pulp at about 20 U/L. The four enzymes present in the mixture act synergistically to accomplish a task, which cannot be achieved by mechanical means.

Benefits Achieved by Adding Proteolytic Enzymes in Fruit Pulp

Reduced solution viscosity, elimination of juice/wine cloudiness, increased juice yield, shorter fermentation period in case of wine making, pectins stabilize the cell debris in a colloidal state; but once pectins are digested, the debris precipitate and are removed by filtration.

Which are the enzymes widely used in the production of fruit juices, beer and distilled alcoholic drinks?

A list of various enzymes widely used in the production of fruit juices, beer and distilled alcoholic drinks are given in following table.

Product	Substrate	Problem	Enzyme(s)	Action/Purpose
Fruit juices, wines	Pectins	Cloudiness	A mixture of: polygalacturonase, pectin esterase, pectin lyase and hemicellulases	Reduced viscosity, no cloudiness cloudiness, increased juice yield, enhanced flavour, shorter fermentation time (wines)
Beer	Starch	Non fermentable	Barley malt + bacterial and fungal α-amylases, β-glucanase	Saccharification; stopped by boiling the 'wort' when 75% starch is saccharified.
	Protein	-	Proteases (neutral)	Hydrolysis of protein and increased fermentation rate (later); especially to obtain high gravity beer.
	Cellulose and barley β-glucans	-	Cellulases	Used to hydrolyse cellulose and barley β-glucans, especially when wheat is added as adjunct.
	Protein (in association with tannin)	Chill-haze on cooling the beer	Papain	Used in post-fermentation stages
	Starch	More total solids in beer	Glucoamylase or fungal α-amylase	Higher degree of saccharification, reduced alcohol and total solids; to produce 'light' beers.
	Dextrins	Non-fermentable more total solids in beer	Genetically engineered yeast (S. cerevisiae)	Utilizes dextrins to produce alcohols; to produce light beers.
Distilled alcoholic drinks	Starch	Non-fermentable	More heat stable bacterial α-amylases	Saccharification of starch present present in the substrate.

Which endogenous enzymes generally used for modification of foods?

Pectic enzymes, amylases, cathepsins, calcium-activated-proteinase, milk proteinase, lipolytic enzymes, thiaminases, phytase, myosin ATPase, lipoxygenase, peroxygenase, peroxidases, ascorbic acid oxidase, antioxidant enzymes (catalase, superoxide dismutase, glutathione peroxidase, sulfhydry oxidase), flavour enzymes, pigment - degrading enzymes etc. are used as endogenous enzymes for modification of various food products.

Which are the factors influencing the enzyme activity?

Temperature, pH, water activity and enzyme activity, effect of electrolytes and ionic strength on enzymes, inactivation of enzymes by shearing, effects of pressure, effects of ionizing radiation, interfacial inactivation etc. factors are responsible for the change in the activity of enzymes in various food and food products.

What is beer?

Beer is an undistilled beverage produced from fermentation of barley malt by yeast, especially Saccharomyces cewrevisiae or Saccharomyces carisbergensis. Other materials rich in starch such as wheat, maize, rice are also added to increase the amount of fermentable sugars and to reduce the relative costs of fermentation; these are called adjuncts.

Barley

Soaking

Malting   Germination at 17°C

Dried at 65°C

Malt

Powdering

Mashing   Adjunct, wheat, rice, maize adding

Mixed with water (70°C pH 5.0)

Mash

Saccharification of Starch

Filteration

Beer wort

Saccharification (ca. 75%)

Saccharification   Boiling with hops

Cooled, inoculated with *Saccharomyces cewrevisiae* or
Saccharomyces carisbergensis

Fermentation   Fermentation at 5-15°C, 5-10 days

Beer

Chilled to 0°C

Stored for 8-10 months

Beer

WHAT IS MEAN BY SINGLE CELL PROTEIN?

Single cell proteins means the proteins produced from single cells from algae, fungi, yeast or bacteria. Single cell proteins may be used directly as human food supplement or it may be used in animal feed to at least replace the currently used protein-rich soybean meal and fish proteins and even cereals, which can be diverted for human consumption.

WHAT SHOULD BE THE CRITERIA OF MICROORGANISMS USED FOR SINGLE CELL PROTEIN PRODUCTION?

It must be non-pathogenic to plants, animals and man; it must have of good nutritional value; easily and cheaply produced on large scale; it should not be toxic; it must be fast growing; it must be easy to separate from medium and to dry etc.

What are the important microorganisms used for single cell proteins production?

Some Important Microorganisms Used in Single Cell Proteins Production

Microorganism	Substrate	Used as	Used commercially
Algae			
Chlorella sp.	CO_2 + sunlight	Feed	Yes, Japan, Taiwan
Scenedesmus acutus	CO_2 + sunlight	-	-
Spirulina maxima	CO_2 + sunlight	Feed	Yes, Mexico
Yeasts			
Candida utilis	Confectionery effluents,	-	Yes, U.K.
	Ethanol,	Food	Yes, USA
	Sulphite liquor	-	Yes, Europe, USA, Russia
C. intermedia	Whey	-	Yes, Vienna
C. krusei	Whey	-	Yes, Kiel process

Contd...

Contd...

C. lipolytica	n-alkanes (C_{10} - C_{23}) + ammonia	-	Yes, Russia
Kluyveromyces fragilis	Whey	Food	Yes, France
Saccharomyces cerevisiae	Molasses	Food	Yes
Fungi			
Chaetomium cellulolyticum	Cellulosic wastes	-	Promising
Fusarium graminearum	Starch hydrolysate	Food	Yes, UK
Paecilomyces varioti	Sulphite liquor	Feed	Yes, Finland
Bacteria			
Brevibacterium sp.	C_1 - C_4 Hydrocarbons	-	Process developed
Methylophilus methylotrophus	Methanol	Feed	Yes, UK

What is the chemical composition of single cell protein produced from various microorganisms?

The nutritional composition of some single cell proteins produced from various microorganisms is given below.

Micro-organism	*Nutrient*		
	Protein	*Fat*	*Ash*
Paecilomyces varioti	55	1	6
Candida utilis	55	5	8
Methylophilus methylotrophus	83	7	9
Spirulina maxima	62	3	2

What are advantages of single cell proteins?

The single cell proteins have several advantages as given below.

1. They can be produced all the year round and are not dependent of the climate except the algal processes.

2. The single cell protein is rich in high quality protein and is rather poor in fats, which is rather desirable.

3. When the substrate used for single cell proteins process is a source of pollution, single cell protein production helps to reduce pollution.

4. The microbes are very fast growing and produce large quantities of single cell proteins from relatively very small area of land.

5. They use low substrates and, in some cases, such substrates which are being wasted and causing pollution to the environment.

6. Mushrooms are considered as delicacy in the human diet.

7. Strains having high biomass yields and a desirable amino acid composition can be easily selected or produced by genetic engineering.

8. Some single cell proteins are good source of vitamins, particularly B-group of vitamins, e.g., yeasts and mushrooms.

9. At present single cell proteins to be the only feasible approach to bridge the gap between requirement and supply of proteins.

10. Single cell proteins are the only source to eliminate protein deficiency among the growing population.

Which are the limiting factors in the biotechnology of food processing?

Following are the some of points explain the limitations of biotechnology in the food processing industry.

1. The very complex nature of many foods, their properties so the establishment of complex is very difficult.

2. The often rather weak relationship between market value and the real intrinsic quality of many foodstuffs (taste, fragrance, texture, appearance etc.).

3. Consumer conservation. Consequently changes have to be brought in rather small and careful steps.

4. Official regulations, more specifically the requirements for the toxicological testing of new products or existing products produced along different bioprocess lines.

5. The shortage of properly trained food engineers with sufficient knowledge in bioprocess technology to be able to fully implement the fruits of biotechnological developments. The relating low emphasis on food engineering and technology in academic circles may be an important factor in this respect.

Which are the factors responsible for making non-availability of nutrients from food and majors should be taken for avoiding such types of difficulties?

Mostly the antinutritional factors at higher level, insecticides, pesticides, binding agents and processing conditions are mostly responsible for making non-availability of the nutrients from various food material to human being. For avoiding such type of difficulties following majors must be taken.

1. Educate farmers, labourers and public on the hazards due to misuse of pesticides, insecticides etc.

2. Need enforcement of regulation governing the use of pesticides and insecticides.

3. Random checking of market samples.

4. To create facilities for food analysis.

5. All agencies connected with agriculture and food commodities must take part in the scheme.

6. It is not enough we produce more food but it must be safe for human consumption.

Why packaging is used preservation of food materials?

In addition to the direct approach to food preservation, such as drying and freezing, other measures such as packaging and quality management tools need to be implemented in the process to avoid contamination or recontamination. Although these measures are not preservation techniques, they can play an important role in producing high-quality safe food. Packaging performs five main functions: product containment, preservation and quality, presentation and convenience, protection and provide storage history.

What is ideal packaging?

There is no such thing as the ideal packaging. Packaging should be such that we could come close to the ideal and the criteria of ideal packaging are given below:

- Strong marketing appeal
- High product visibility
- Zero toxicity
- Ability of moisture and gas control
- Low cost and availability
- Stable performance over a large temperature range
- Suitable mechanical strength
- Ability to include proper labeling
- Protection from loss of flavour and odours
- Controlled transmission of required or unwanted gases
- Easy machine handling and suitable friction coefficient

- Closure characteristics, such as opening, sealing and resealing, pouring
- Resistance of migration or leaching from package

CODEX ALIMENTARIUS AND HACCP

Good manufacturing practice (GMP) and good hygiene practice (GHP) describe the basic measures that have to be applied during production, processing, handling and distribution, storage, sale, preparation and use. The GMP and GHP developed by Codex Alimentarius Commission on food hygiene and the food industry. General requirements are as follows:

1. The hygienic design and construction of food manufacturing premises
2. The hygienic design, construction and use of proper machinery
3. Cleaning and disinfections procedures
4. The microbial quality of raw foods
5. The hygienic operation of each process step
7. The hygienic of personnel and their training.

SEVEN STEPS OF THE HACCP

1. **Hazard Analysis :** Identify critical food-production steps where hazards might occur, assess their severity and human health risks, and determine preventative measures.

2. **Critical Control Points :** Identify critical control points in the process at which the potential hazard can be controlled or eliminated.

3. **Specification of Critical Limits :** Establish control measures and set up criteria to measure control at those critical points.

4. **Monitoring :** Monitor critical control points by establishing procedures for how the critical measures will be monitored.

5. **Corrective Actions :** Take corrective action when the criteria are not being met.

6. **Verification :** Routinely check the system for accuracy to verify that it is functioning properly and consistently.

7. **Documentation :** Establish effective record-keeping procedures that document and provide a historical record of the facility's food safety performance.

STATUTORY PROVISIONS FOR QUALITY CONTROL IN INDIA

These provisions are come in to existence for several reasons.

- To maintain the quality of food produced in the country.
- To prevent exploitation of the consumer by the sellers.
- To safeguard the health of the consumers.
- To establish criteria for quality of food products, since more and more foods are processed, rather than in natural forms. This has resulted in the inability of the consumer to identify the quality of the contents that could be identified easily.

  1. Prevention of Food Adulteration Act 1954 and Rules 1955 (PFA Act).
  2. Fruit Products Order 1955 (FPO) Regulates the manufacture, storage and sale of fruit and vegetable products.
  3. Agricultural Produce (Grading and Marketing) Act 1937 [AGMARK]
  4. Sugar (Control) Order 1956.
  5. Vegetable Oil Products (Control) Order 1947; The Solvent Extracted Oil; Deoiled Meal and Edible Flour (Control) Order 1967; Vanaspati Control Order 1975.
  6. Meat Food Products (Control) Order 1975.
  7. Rice Milling Industry (regulation) Act 1958 and Regulation and Licensing Rules 1976.
  8. Export (Quality Control and Inspection) Act 1963 and Rules 1964.
  9. Insecticide Act 1968.
  10. Standards of Weights and Measures Act 1976.
  11. State Licensing Order Governing Grain Dealers.
  12. The consumer Protection Act 1986.

FOOD STANDARDIZATION AND REGULATORY AGENCIES IN INDIA

Following are the various agencies working in India for Food Standardization and Regulations.

1. Central Committee for Food Standards
2. Central and State Food Departments
3. State Food Laboratories/Food and Drug Administration
4. Bureau of Indian Standards
5. Food corporation of India
6. Army supply corps
7. Central insecticide Board

FOOD FADS AND FALLACIES

Fallacious beliefs concerning foods and their usefulness to the body are not new.

- That fruits especially citrus and tomato are too acid to be handled by the body: The acids in fruits are organic acids. After absorption they are oxidized to carbon dioxide and water and readily eliminated from body. Fruits are base forming with the exception of prunes, plums and cranberries, which contain organic acids (benzoic and quinic) that the body cannot oxidize.

- That garlic cure high blood pressure: No evidence.

- That beets build blood: Not essential for blood formation. Protein and iron are the chief blood-building constituents as well as cu; folacin and vitamin B_{12} are essential.

- That foods cooked in aluminum utensils will cause cancer: No evidence or not true. Small amount can exerts from body.

- That the milk and orange juice or other citrus fruit, milk and fish combinations are poisonous: The belief is entirely errorness. Many with no ill effects eat such food combinations daily.

- That raw cucumbers without salts are poisonous: No evidence.

- That a good way to diet is to skip breakfast: It lowers physical efficiency.

- That honey is not fattening: No particular food is fattening.

- That meat gives you strength: Meat has a high satiety value. The extractives give a flavour that is enjoyed and the protein is effective in delaying the onset of hunger sensations.

- That fruit juices do not contribute calories to the diet: Fruit juices contribute calories to the diet as follows. Per cup: Lime 65, Lemon 60, Orange 110, Pineapple 135, Prune 200.

- That toast has fewer calories than bread: Same value in calories.

- That vegetable fats and oils can be used in any quantity and are not fattening: Vegetable fats and oils are high in caloric value, and are comparable to animal fats.

- That adults need no milk: It is difficult to supply the recommended calcium allowance of adults without milk. In addition, dairy products provide in significant amounts protein of high quality, phosphorus, vitamin A and riboflavin.

- That skim milk has little nutritive value: It remains a superior food, even though low in fat and vitamin A content.

- That pork liver is less nutritious than beef or calves liver: These all have same energy value and protein content. Pork liver is higher in iron and thiamine content and lower in vitamin A.

- That white-shelled eggs are more nutritious than brown: Nutritive value is not related to colour of the shell, which is determined by the breed of the hen.

- That water is fattening: Water has no calorific value and therefore cannot be converted to body fat.

TOTAL QUALITY MANAGEMENT FOR NATIONAL AND INTERNATIONAL MARKET

Total quality management (TQM) emphasis is on creating an organizational culture that involves extensive participation, an emphasis on teams and teamwork, cooperation between units, generation of valid data, and continuous learning, TQM is highly congruent with organizational development approaches and values.

Most companies are interested to improve quality of their products and services through TQM. Total quality management approach focuses on trying to meet customer expectations or delighting the customer. All quality improvement initiatives must begin with an understanding of customer perceptions and needs. TQM is an organizational strategy with techniques that deliver quality products and/or services to customer and achieves total customer satisfaction. It should be customer, which comes back and not the product.

Under TQM not only the "Customer is King" but so are internal customers such as co-workers or other departments. TQM becomes the dominant culture of the organization. Some core values of every one involved in TQM are as follows:

1. Make it right for the customer at any cost.

2. Customer is always right.

3. Internal customers are as important as external customers.

4. Respond to customer inquiry or complaint by the end of the day.

5. Answer the phone bell within two rings.

6. Not only meet customer expectations but delight customers in the process.

7. Teamwork and cooperation are important.

8. Every one is involved in quality effort.

9. Respond to every employee's suggestion for quality improvement.

10. Always strive for continuous improvement. Never be satisfied with level of quality.

Concept of Total Quality Management

Total	*Quality*	*Management/Control*
1. Covers all functional areas	1. Conformance to customer's needs expectations, quality service	1. Effective utilization of: • Mean • Machines • Materials • Money • Time
2. Covers all employees at all levels. It is employee-centered.	2. Fitness for use	2. Work towards continuous improvement in all spheres and activities of an organization.
3. Covers all others: • Suppliers • Customers	3. Customer satisfaction	3. Who have a stake in the organization

STEPS THAT IDENTIFY ORGANIZATION PROCESS TOWARDS TQM

Quality is not absolute but continuously changing the perception.

1. Awareness
2. Involvement
3. Commitment
4. Ownership

FOUR STREAMS OF THE TOTAL QUALITY MANAGEMENT

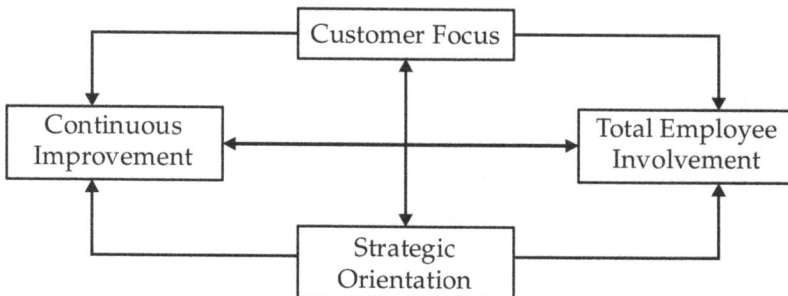

PRINCIPLES, PRACTICES AND TECHNIQUES OF TOTAL QUALITY MANAGEMENT

The principles, practices and techniques of total quality managements are given in details in table.

There are there innovative techniques of total quality management.

1. Reengineering
2. Bench-marking and
3. Empowerment.

1. **Reengineering :** Reengineering is the fundamental rethinking and radical redesigns of business processes to achieve improvements in performance, such as cost, quality service. Objective is to eliminate inefficiencies and increase productivity and performance.

2. **Benchmarking :** Bench marketing is the process of comparing work and service methods against the best practices for the purpose of identifying changes that will result in higher quality out put. It is finding best practice, measuring the gap between those and ones own and bridging it.

3. **Empowerment :** Empowerment is the authority to make decisions within one's area of operations without having to get approval from any one else. There are two characteristics that make empowerment unique:

   1. Personnel are encouraged to use their initiative and
   2. The employees are given not just authority but resources as well.

Basic conditions for empowerment:

1. Participation
2. Innovation
3. Access to information
4. Accountability

QUALITY CIRCLES AND TOTAL QUALITY MANAGEMENT

Quality circle is a form of group problem solving, participation at workplace and goal setting with primary focus on maintaining and enhancing product or service quality. Quality circle inspires cooperation, teamwork, problem-solving capabilities, safety awareness and cost reduction. These promote harmonious supervisor-worker relationships and leadership development. Quality circles have been over taken by a more radical approach called total quality management. This is a holistic approach, which fits the organized development definition of planned whole organization change.

Comparison of Quality Circles and Total Quality Management

Feature	Quality circles	Total quality management
• Choice of membership	• Voluntary	• Compulsory
• Structure	• Add-on	• Integrated
• Direction	• Bottom-up	• Top-down
• Aims	• In departments	• Company wide
	• Employee relations and work improvements	• Quality culture and quality performance

QUALITY CONTROL IN FOOD SERVICE

There are several points where food products quality can be control.

1. Receiving and inspection control
2. Storage and issuing control
3. Pre-cooking quality control
4. Cooking quality control
5. Post-cooking quality control
6. Desserts and baked products control
7. Non-alcoholic beverages control
8. Food spoilage and sanitation control
9. Water quality and ware washing control
10. Quality control of vending equipment
11. Energy control and conservation

1. SCOPE OF QUALITY CONTROL IN FOOD SCIENCE

What is quality control?

Quality control, or quality assurance is an activity procedure, method or programme that will ensure the maintenance and continuity of specifications and standards of a product within prescribed tolerances during all stages of handling, processing, preparation and packaging and will further ensure that all the original and desirable characteristics are sustained during storage, processing or preparation and will remain unaltered until consumed.

The current awareness by management that if a quality control programme is not carried out to the fullest extent, especially where efficiency food items are the major source of revenue, growth will suffer and failure many result.

THE CONSUMER'S INTERPRETATION OF QUALITY

(a)   Personal preferences

1.   Liked

2.   Disliked

3.   Excellent

4.   Superior

5.   Great

6.   Good

Many factors influence the consumer's decision, such as habit, locality, ethnic characteristics, advertising, gimmicked sales promotions and price.

In addition to these psychological factors, positive sensory stimulation plays an important role in establishing quality parameters. These include an appealing favours, a pleasing mouth feel or texture, an attractive natural colour or appearance, general palatability, product consistency and to many customers, the nutritional value of the food.

(b)   The technical interpretation of quality

Analyst or technologist refers to quality

(a)   Chemical or physical measurements

(b)   Consumers acceptance/satisfaction

(c)   Management's interpretation of quality

Management equates quality with certain economic factors, such as the cost of the product, profits generated and consumer acceptance within the intended selling price range. It wills ensure/help for a healthy long-range growth.

Factors Affecting Food Quality

1.   Poor sanitation.

2.   Faulty handling.

3.   Malfunctioning equipment.

4.   In correct preparation.

5.   Carelessness.

6.   Low Quality raw material.

7.   Location of the factory/Food industry.

Prime Factors responsible for Significant Quality Changes.

1.   Spoilage due to microbiological, biochemical, physical or chemical factors.

2. Adverse or incompatible water conditions.

3. Poor sanitation and ineffective ware washing.

4. Improper and incorrect pre-cooking, cooking and post-cooking methods.

5. Incorrect temperatures.

6. Incorrect timing.

7. Wrong formulations, stemming from incorrect weight of the food or its components.

8. Poor machine maintenance programme.

9. Presence of vermin and pesticides.

10. Poor packaging.

Coffee is one of the example in which following Several Factors Affect the Products Quality

1. Size of green coffee beans.

2. Bulk density of green coffee beans.

3. Chemical composition of green coffees.

4. Blend of green coffee.

5. Roasting technique.

6. Degree of roast.

7. Colour of roast.

8. Chemical composition of roasted coffee.

9. Bulk density of roasted coffee.

10. Particle size distribution of ground coffee.

11. Water composition.

12. Temperature of water.

13. Volume of water.

14. Weight of coffee.

15. Degree of contact between water and coffee.

16. Wettability.

17. Time of contact.

18. Separation of beverage from grounds.

19. Clarity of beverage-freedom from sediment.

20. Length of holding period before drinking.

21. Temperature during holding period.

22. Mixing of finished brew.

23. Methods of serving.

24. Cleanliness of brewing and serving equipment.

Fruit and vegetable development starts with formation of an edible-fruit setting, seedling emergence, tuber development or stalk development and ends with loss of edible character through physiological deterioration, development of fibrous character, or spoilage through microbiological intervention.

In practice one or more of following methods for human consumption or processing into various food products determine fruits and vegetables maturity.

(a) Computation of days from bloom to harvest.

(b) Measurement of heat units.

(c) Visual means–skin colour, persistence or drying of parts of the plant, fullness of fruit etc.

(d) Physical methods—ease of separation, pressure test, density grading etc.

(e) Chemical method-total solids, sugars, acid, sugar-acid ratio, starch content etc.

Maturity and Quality Grades for Fruits and Vegetables

Fruit/Vegetable	Maturity and Quality Grades
Apple	Skin and flesh colour, flesh firmness, sugars, acid, starch accumulation from full bloom to harvest.
Apricot	Dependent on intended purpose, soluble solids, flesh firmness, light transmission properties.
Banana	Inflorences drops, sugar accumulation, colour change to pink, ribs soften.
Ber	Sugar accumulation, colour change to pink, low acidity.
Citrus	Fruit colour, taste, soluble solids/acid ratio.
Custard apple	Sugar accumulation, eyes opening, colour becomes pink.
Fig	Sugar accumulation, colour change.
Grapes	Sugar accumulation, loss of acids, berry softening, and skin colouration, ripening with varietal flavour or odour development.
Guava	Fully mature, colour changes to dull pink, increase softness.
Mango	Fully mature, colour change mostly pink, naturally fruit drops.
Pear	Flesh firmness, ease of spur separation, harvested hard green, ripened off the tree; colour, texture, sweetness, typical varietal flavour development.

Contd...

Contd...

Fruit/Vegetable	Maturity and Quality Grades
Pomegranate	Become hard, colour changes, dryness.
Sapota	Sugar accumulation, colour change.
Strawberry	Fully red or pink.
Beans	Harvest at 14-18 days from full bloom, size, fibrousness
Broccoli	Firm head, closed florets, colour, shape, lack of yellowing
Cabbage	Head firmness, appropriate colour (green, purple), and size.
Cauliflower	Size, curds tight and compact, colour.
Carrot	Size and colour suitability for intended use.
Chilli	Colour (green, red, and yellow).
Methi	Before flowering, tenderness of leaves.
Onion	Top bending, size.
Potatoes	Harvesting at 90-120 days after planting
Peas	Tenderness, size.
Pumpkin	Size, shape and colour.
Radish	Size and cleanliness, suitability for use.
Spinach	Colour, freshness, full growth.
Tomatoes	Mature green, pink, red etc.

Quality Control Management in Food Processing Industry

Food quality control is generally defined as the regulation by law of food manufacture, distribution and sale, in order to prevent health hazards and fraud to the consumer. Thus, it becomes a criminal offence to sell (deliberately or in any other way), adulterated, filthy contaminated food. Food is defined under the law as any article used for consumption as a food or drink by human or animal.

There are three main aspects to the application of food quality control: 1. Moral, 2. Commercial and 3. Legal. The legal viewpoint demands that the quality of products conforms to national and international strands. Fruits and vegetables processing industries produce very large quantities of products, which are intended for consumption, often on a daily basis, by the population at large.

The control of food quality by law leads to:

1. Improved quality of product.

2. Achievement of greater consumer satisfaction.

3. The promotion of quality consciousness.

4. Increased consumption and sales.

5. Employment opportunities for scientific and technical personal.

6. Avoidance of controversy and litigation in marketing at the national and international level.

7. Promotion of national and international trade.

8. Provision of the means for the intelligent comparison of prices in relation to quality and grade.

9. Greater confidence in the minds of consumers.

Quality control within a food manufacturing industry demands constant vigilance at all stages in processing, so that any necessary adjustments can be made at the appropriate time.

The specific responsibilities of quality control assigned to a department or to an individual include following points.

1. Standardizing procedure for sampling and examining raw materials.

2. Development of test procedures.

3. Establishment and implementation of quality standards for fresh and processed products.

4. Setting up preventive quality control methods for in-plant liaison between manufacturing section and test laboratories.

5. Examination of finished products.

6. Storage controls.

7. Recording and reporting.

8. Special problems, including attendance to consumer complaints by locating their causes and eliminating them.

9. Research and development into new products and their packaging.

THE SEQUENCE OF OPERATIONS IN QUALITY CONTROL

1. Raw material control.

2. Process control or the control of the manufacturing process.

3. Production inspection, including the inspection of the finished product, packaging and storage.

4. Sensory evaluation or evaluation of the acceptability of the final product.

FOOD SAFETY MANAGEMENT BY HACCP SYSTEM

The acronym HACCP stands for Hazard Analysis and Critical Control Point, which is a prevention-based food safety management system. Essentially

HACCP is a management system that identifies and monitors specific food borne hazards, biological, chemical or physical properties, which can adversely affect the safety of the food product. Seven principles have been developed which provide guidance on the development of an effective management of HACCP plan. Food and Drug Administration is recommending the implementation of HACCP in food establishments because it is management system of preventive controls that is the most effective and efficient way to ensure that food products are safe.

The National Advisory Committee on Microbial Criteria of Food (NACMCF) has developed seven widely accepted HACCP principles that can explain this process in great detail.

1. **Principle 1:** Hazard Analysis; Likelihood that the hazard will occur and the severity if it does occur.

2. **Principle 2 :** Identify the Critical Control Points (CCP); Points in food preparation that may be CCPs include cooking, chilling, specific sanitation procedures, product formulation control, prevention of cross contamination, and certain aspects of employee and environmental hygiene.

3. **Principle 3 :** Establish Critical Limits for Preventive Measure; e.g., Temperature, time, physical dimensions, a_w, pH, and available chlorine etc.,

4. **Principle 4 :** Establish Products to Monitor CCPs; (a) Observations and measurements, (b) Continuous monitoring, (c) Monitoring procedures

5. **Principle 5 :** Establish the corrective action to be taken when monitoring shows that a critical limit had been exceeded. A. Purpose of Corrective Action Plan, B. Aspects of Corrective Action Plan.

6. **Principle 6 :** Establish Effective Record Keeping Systems that Document the HACCP System.

7. **Principle 7 :** Establish Procedures to Verify that the HACCP System in Working. A. Establishing Verification Procedures, B. Examples of HACCP Plan verification activity, C. Training and Knowledge gaining, 1. Focus and objective, 2. Reinforcement etc. FAO/WHO and ISO standards for specific fruit and vegetable products are listed in tables.

PROSPECTS FOR QUALITY CONTROL SERVICES

Though quality is essential to food processing and the successful marketing of the products but small and medium-scale food industries, especially young ones are not able to set up quality control laboratories. Prospects are quite good for chemists and qualified technical personnel to use their professional training to set up private laboratories providing quality control and other analytical services for local food processing industries. Capital is needed initially to purchase laboratory equipments, glassware and chemicals. But such costs may eventually be recovered as demand for the laboratory services grows. This is likely to happen as food-processing enterprises compete and as quality strands come to be enforced by law.

Principles, Practices and Techniques of Total Quality Management

Item	Customer focus	Continuous Improvement	Teamwork
Principles continuous	Paramount improvement of providing products and services that fulfills customers needs; requires organization-wide focus on customers.	Consistent customer satisfaction can be attained only through relentless improvement of processes that create products and services.	Customer focus and improvement are best achieved by collaboration through in organization as well as with customers and suppliers.
Practices	• Direct customer contact, collecting information about customer needs. • Using information to be sign and deliver products and services.	• Process analysis. • Re-engineering problem solving.	• Search for arrangements that benefit all units involved in a process. • Formation of various types of teams. • Group skills training.
Techniques	• Customer surveys and focus groups. • Quality function deployment (translates customer information into product specifications).	• Flowcharts. • Pareto analysis. • Statistical process control. • Fishbone diagrams.	• Organizational development method such as the nominal group technique. • Team building methods, (e.g., role clarification and group feedback).

FPO and ISO Specifications for Various Fruits and Vegetable Products for National and International Markets

Fruit/Vegetable Products	FPO Specifications
Canned Fruits	
Drained weight	Not less than 50%, For berry 40%
Texture	Free from disintegration, damage from bruises and uniformly prepared
Added colour	Erythrosine in canned cherries
Organoleptic quality	Characteristic taste
Preservatives	Not permitted

Contd...

Fruit/Vegetable Products	FPO Specifications
Canned vegetables	
Drained weight	Not less than 55%, canned tomato 50%
Texture	Free from disintegration and damage from bruises
Colour	Pink discolouration test for leucoanthocyanins
Added colour	Not permitted except for processed peas
Salt preservatives	Not permitted
Fruit juices and Concentrates	
Juice content unsweetened	Natural 100%
Sweetened	85%
Concentrate orange juice	40%
Others except from tomato juice	Natural, 100%
Total soluble solids	Natural, not less than 10%
Acidity as anhydrous citric acid	Lime juice not less than 5.0% Lemon juice not less than 4.0% Others not greater than 3.5%
Synthetic sweetening agents	Not permitted
Preservatives Sulphur dioxide	Not more than 350 ppm
Benzoic acid	Not more than 600 ppm
Added colour	Permitted colours
Incubation test	No sign of bacterial growth on incubation at 37°C for 7 days
Fruit Nectar	
Juice content	Pineapple and orange not less than 40%, others not les than 20%
Total soluble solids	Not less than 15%
Organoleptic quality	Free from objectionable taints and odours
Fermentation test	No sign of bacterial growth on incubation at 37°C for 7 days
Tin content	Not more than 250 ppm
Mango pulp	
Natural juice content	Original and should pas through 1.5 mm mesh
TSS	Not less than 12%
Sweetened Juice content	Original
TSS	Not less than 15%
Acidity as citric acid	0.3%
Pulp characteristics	Should pass through 1.5 mm mesh, Be free from cooked flavour, black specs extraneous matter like portion of skin fibrous matter, larvae and insect or its fragments

Contd...

Fruit/Vegetable Products	*FPO Specifications*
Incubation test	No sign of bacterial growth when incubated at 37°C.
Can pressure	Negative at sea level
Soft drinks Ready-to-serve Beverage-Natural	
Juice content Lime	Not less than 5%
Others	Not less than 10%
Total soluble solids	Not less than 10%
Preservatives	
Sulphur dioxide	Not more than 70 ppm
Benzoic acid	Not more than 150 ppm
Synthetic sweetening agent	Not permitted
Added colour	Permitted colours
Ready-to-Serve Beverages Synthetic	
Total soluble solids	Not less than 8%
Total sugars	Not less than 5%
Saccharine	Not more than 100 ppm
Caffeine	Not more than 200 ppm
Emulsifying and stabilizing agents	Permitted
Edible gums and gelation	Permitted
Squash, Crush, Fruit Syrup, Cordial and Barley Water	
Juice content	Not less than 25%
Total soluble solids	
Squash	Not less than 40%
Crush	Not less than 55%
Fruit syrup	Not less than 65%
Cordial	Not less than 30%
Barley water	Not less than 30%
Acidity as anhydrous citric acid	Squash, crush, fruit syrup and cordial not more than 3.5%, barley water not more than 2.5%
Barley starch in barley water	Not more than 0.25%
Added colour	Permitted colours
Preservative	
Sulphur dioxide	Not more than 350 ppm
Benzoic acid	Not more than 600 ppm
Synthetic sweetening agent	Not permitted
Clarity in cordial	Clear, free from pulp and other cellular
Fermentation test	Negative at 37°C.
Organoleptic quality	Free from objectionable taints and flavours

Contd...

Fruit/Vegetable Products	*FPO Specifications*
Tomato Juice	
Appearance	Free from pieces of skin, seeds, bits, of coarse tissue and extraneous matter
Total soluble solids	Not less than 5%
Salt	Not more than 1.5% w/w
Added colour	Permitted colours
Organoleptic quality	Characteristic and be free from burnt or other objectionable flavours
Incubation test	No sign of fermentation when incubated at 37°C for 7 days
Mould count	Not more than 30% of the fields examined
Rot fragment	Not permitted
Insect fragment and other filth	Not permitted
Tomato Puree and Paste	
Total soluble solids, free of salt	Tomato puree not less than 9%, w/w Tomato paste not less than 25% w/w
Added colour	Permitted colours
Preservatives	Benzoic acid not more than 250 ppm
Incubation test	No sign of fermentation when incubated at 37°C for 7 days
Mould count	Not more than 60% of fields examined
Appearance	Free from skins and seeds
Organoleptic quality	Characteristic and free from burnt or objectionable flavour
Rot fragments	Not permitted
Insect fragments and filth	Not permitted
Tomato Soup	
Total soluble solids, free of salt	Not less than 7% w/w
Salt	Not more than 1.5% w/w
Added colour	Permitted colours
Incubation test	No sign of fermentation when incubated at 37°C for 7 days
Organoleptic quality	Free from burnt or objectionable flavour
Mould count	Not more than 30% of fields examined
Insect fragments and filth	Not permitted
Vegetable Soup	
Added colour	Permitted colours

Contd...

Fruit/Vegetable Products	FPO Specifications
Consistency	Uniform
Chemical preservatives	Not permitted
Rot fragments	Not permitted
Insect fragments and filth	Not permitted
Incubation test	No sign of fermentation when incubated at 37 0C for 7 days
Organoleptic quality	Characteristic of the vegetable used
Tomato Ketchup and Sauce	
Total soluble solids	Not less than 25% w/w
Acidity	Not less than 1.0%
Added colour	Not permitted
Preservative	Benzoic acid not more than 750 ppm
Mould count	Not more than 40% of field examined
Yeast and spore count	Not more than 125 per 1/60 cmm
Bacteria	Not more than 100 million per cc
Rot fragment	Not permitted
Insect fragments	Not permitted
Heavy extraneous matter	Not permitted
Appearance	Free from skins and seeds
Organoleptic quality	Characteristics and free from burnt or other objectionable flavors
Incubation test	No sign of fermentation when incubated at 37°C for 7 days
Jam, Jelly, Marmalade and Fruit Cheese	
Fruit content	All jams, jellies, marmalades and cheese not less than 45%, Raspberry and Straw berry Jam not less than 25%
Total soluble solids	Jams and Cheese not less than 68% w/w Jelly and Marmalade not less than 65%
Preservatives Sulphur dioxide Benzoic acid	Not more than 40 ppm Not more than 200 ppm
Synthetic sweetening agents	Not permitted
Added colour	Permitted colours
Mould growth	Absent
Fermentation test	Negative pressure at sea-level
Crystallization	Absent

Contd...

Fruit/Vegetable Products	FPO Specifications
Fruit Chutney	
Fruit content	Not less than 40% any fruit when calculated with raisins and dry fruits, if used in excess of 5% or 40% fruit content in mango chutneys or others shall be declared on the label.
Total soluble solids	Not less than 50% w/w
Acidity	Not less than 2.0%
Alum	Not permitted
Preservative Sulphur dioxide Benzoic acid	Not more than 100 ppm Not more than 250 ppm
Mould count	Not more than 40% of fields examined
Rot fragments	Absent
Insect fragments	Absent
Incubation test	Negative at 28-30°C and 37°C
Sign of fermentation	Negative
Pickles in Vinegar	
Drained weight	Not less than 67%
Drained liquid	One-third of the total content
Total acidity	Not less than 2% in the liquid
Clarity	Reasonably free from sediment
Preservatives	Not permitted
Alum	Not permitted
Mineral acids	Not permitted
Added colour	Not permitted
Added copper	Not permitted
Heavy metals Lead Copper Zinc Arsenic Tin	Not more than 4 ppm Not more than 10 ppm Not more than 5 ppm Not more than 0.5 ppm Not more than 200 ppm
Fungal attack, insect fragments	Absent
Extraneous matter	Absent
Fermentation test	Negative
Oil Pickles	
Added copper	Not permitted
Mineral acids	Not permitted

Contd...

Fruit/Vegetable Products	FPO Specifications
Alum	Not permitted
Preservatives	
Sulphur dioxide	Not more than 100 ppm
Benzoic acid	Not more than 250 ppm
Insect fragments	Absent
Heavy extraneous matter	Absent
Sun-dried and Dehydrated Vegetables	
Appearance	Free from stalks, peels, stems and extraneous leaves
Rehydration ratio	Reconstitute to original shape and quality by boiling for 15 - 60 min.
Sulphur dioxide	Not more than 2000 ppm
Acid insoluble ash	Not more than 0.5%
Visible mould	Absent
Insect or larvae	Absent
Fruit, Fruit Pulp and Fruit Juice for Further Processing	
Preservative sulphur dioxide cherries	Not more than 5000 ppm
Strawberry and raspberries other fruits	Not more than 2000 ppm
Insect fragments and other extraneous matter	Absent
Tamarind Concentrate	
Pressure	Negative at sea level
Acid as tartaric	Not less than 9%
TSS	Not less than 65%
Ash insoluble in HCl	Not more than 0.8%
Flavour	Free from burnt or other objectionable flavour
Moulds	Insects and their fragments, rodent contamination Absent
Extraneous matter	Free from stalk and fibrous matter

Fruit	Maturity ISO Quality Parameters		
	TSS (%)	Acidity (%)	Juice (%)
Mandarin	12 - 14	0.4	33
Grape fruit	11	-	35

Contd...

Contd...

Lemon	-	-	25
Limes	-	6	25
Pears	12	-	-
Grapes	20	-	-
Mango	18	0.32	-

FPO Specifications for Bottled or Canned Fruits and Vegetables

Product	Specifications
Bottled or canned fruit	1. Headspace in the can shall not be more than 1.6 cm.
	2. Te drained weight of the fruit shall not be less than 50 per cent and the fruit should be firm.
	3. No preservative shall be added.
	4. No artificial colour shall be present, except in case of peas where permitted colour may be added.
	5. The can shall not show any positive pressure at sea level and shall not show any sign of bacterial growth when incubated at 37°C for a week.
Bottled or canned vegetable	1. Headspace in the can shall not be more than 1.6 cm.
	2. The drained weight of the vegetable shall not be less than 55 per cent (50% in the case of tomatoes).
	3. No preservative shall be added.
	4. No artificial colour shall be present, except in case of peas where permitted colour may be added.
	5. The can shall not show any positive pressure at sea level and shall not show any sign of bacterial growth incubated at 37°C for a week.

FAO/WHO Codex Alimentarius Commission International Maximum Limits for Pesticide Residue in Fruits Codex Tolerance Limits in mg/kg.

Fruit	Carbaryl	Ethion	Folpet	Malathion	Ortho-phenylphenol	Phosphamidon	Trichlorfon	Lindane
Apple	5	2	10	2	-	0.5	0.1	0.5
Banana	5	-	-	8	-	-	-	-
Citrus	7	2	10	-	10	0.4	0.1	-
Grape	5	2	2.5	-	3	-	-	3
Pineapple	-	-	0.01	-	10	-	-	-
Strawberry	7	2	20	1	-	0.2	-	3
Watermelon	-	-	2	-	-	0.1	-	-

FAO/WHO Codex Alimentarius Commission International Maximum Limits for Pesticide Residue in Vegetables Codex Tolerance Limits in mg/kg.

Vegetable	Aldrin and Dieldrin	Carbaryl	Chlorofen-vinphos	Malathion	Phosph-amidon	Trichlorfon
Bean	-	5	-	-	0.2	-
Broccoli	0.1	-	0.05	5.0	0.2	-
Cabbage	0.1	-	0.05	8.0	0.2	-
Carrot	0.1	2	0.4	-	0.2	-
Cauliflower	0.1	-	0.1	0.5	-	0.2
Cucumber	0.1	3	-	-	-	-
Eggplant	0.1	5	0.05	0.5	-	-
Okra	0.1	10	-	-	-	-
Onion	0.1	-	0.05	-	-	-
Potato	0.1	-	0.05	-	-	-
Radish	0.1	2	0.1	-	-	-
Spinach	-	-	-	-	-	0.2
Tomato	-	5	0.1	3.0	0.01	-

7

HUMAN NUTRITION AND DIETETICS

General Human Nutrition and Dietetics

1. Improvements in the health of population depend largely on social and economic factors such as diet, hygiene, immunization and changes in the roles and status of women.

2. Following the cholera epidemic of the 1830s, there was a growing acceptance of a governmental role in public health. This was reinforced by the new understanding provided by the germ theory, of the causes of diseases and by the identification of widespread malnutrition as a factor underlying ill health and poor physical and mental performance.

3. The discovery of vitamins growth promoting diets for children and the malnutrition of the poverty underlined the importance of the diet for the nation's health.

4. During World War II, for the first and only time a coherent nutrition and food strategy formed major part of UK Government policy. This wartime approach won worldwide acclaim.

5. From the 1950s on words, concerns about nutritional deficiency switched to the developing countries. Meanwhile in Western Europe and the United States, a revolution in farming methods and the development of "agribusiness" led to excess production of almost every food, specially cereals, dairy and meat products.

6. In the second and third worlds there are emerging nutritional problems related to the pressure to switch to Western diets, the collapse of the command economics of Eastern Europe and urbanization in developing countries.

7. During the 1980s, the UK, it has become evident that the traditional control for ensuring the toxicological and microbiological safety of foods have been inadequate. However, internationally while food safety is seen as important the long term issues of nutrition and health are given less attention.

8. In Western Europe and North America increased affluence the development of the food industry and of supermarket retailing mean that processed food forms the major part of people's diets with consequent increases in fat, sugar and salt intake.

9. The modern diet brings new concerns about health and ethics of production methods.

MALNUTRITION, KEY POINTS VINPHOS

1. Type I and type II deficiency syndrome are fundamentally different types of response to deprivation of the single nutrients. Deficiencies of type II nutrients are the cause of chronic malnutrition in 50% of the world's children and are responsible for unrecognized health problems in many others.

2. Methods of assessing malnutrition need to identify the most at risk are simple to apply and be internationally accepted. In children, height for age, weight for age and in upper arm circumference (MUAC) is most useful and in adults body mass index (BMI) and MUAC are employed.

3. In severe malnutrition there are three main clinical syndromes, marasmus, kwashiorkor and dwarfism which often coexist in the same individual with primary malnutrition.

4. Clinical features of the malnutrition also include edema, abdominal swelling, and hepatomegaly, anorexia and changes in skin, hair, checks and bone. Gynaecomastia sometimes occurs. Hyporolamia occurs in severe malnutrition.

5. The loss of weight leads to a reduction in the absolute nutritional requirements. However, there is also a relative reduction in requirement as the cells of the body do less work, with a reduction in the functional capacity of the organs. There are reductions in cardiac output, intestinal motility, the activity of the sodium pump and intensity of turnover. The patient tends to be come poikilothermic and there are changes in body composition and in hormonal balance, the inflammatory and immune responses are blunted or absent.

6. Oedematous malnutrition is an acute illness. It is associated with a variety of debilitating diseases, and is metabolically different from marasmus.

7. In acute phase management of malnutrition, immediate threats to life are addressed and treatment is begun which aims at reversing the physiological changes without overloading lover. In the intermediate phase, energy and nutrients are given in amounts marginally above the patient's maintaince requirement to correct metabolic abnormalities; appetite is used as an indicator. In the rehabilitation phase, the patient's body is ready to make more tissue and the patient is fed to appetite, attention should also be given to emotional and physiological needs. Towards the end of this phase and up to following discharge, there should be education of the parents and caretakers with regard to continuing the rehabilitation and preventing recurrence.

8. The presence of children with severe malnutrition may suggest a need for a community wide programme of nutritional and health education.

9. The principles of the treatment for malnourished children's with HIV are exactly the same as for those without the infection.

OBESITY

1. In obesity body fats stores are so enlarged as to impair health. A generally accepted definition of obesity is a Quetelets Index (QI) or body mass index (BMI) of greater than 30 kg/m².

2. In the UK the prevalence of obesity among both men and women has more than doubled since 1980. A similar trend is found throughout developed and third world countries, although it arises from widely differing baseline levels.

3. Obesity reduces the sensitivity of the tissues to the action of insulin and thus increases the risk of non-insulin dependent diabetics, heart disease, hypertension and stroke. It is also associated with increased risks for some cancers gall stems, some reproductive disorders and obstructive sleep aphoea. Obese people are relatively disadvantaged psychologically, socially and economically; the psychological wellbeing of children and adolescents is particularly affected.

4. All the ill effects of obesity decreases with weight loss except for gallstone formation.

5. When there is long term excess of energy input over output, new energy equilibrium is eventually reached, because of increases in dietary thermo genesis and metabolic rate, at a higher weight than previously.

6. Estimates concerning the genetic component if obesity range from 5% to 70%. It is probable that predisposition to obesity arises from interaction between the environment and many different genes.

7. In activity can be reasonably assumed to be an important cause of obesity and impaired glucose tolerance. Social factors such as low educational level also contribute to a predisposition to obesity. Endocrine defects are very rarely a primary cause. The balance, of the micronutrients in the diet cannot cause obesity; it is the differences between the macronutrients in the energy density and palpability that makes diet more likely to stimulate a greater energy intake.

8. Before advice is given to obese patient, essential background factors should be considered, and also the appropriate rate of weight loss should be agreed with the individual patient.

9. It is important that the healthcare professional maintains the relationship with the obese patients which produce benefit for the latter. The treatment

objectives are to help the patient to achieve the weight within the desirable range to maintain this weight and where necessary to attempt to restore the patient self esteem.

10.   Reducing diets should be within the range of 800-1500 kcal/day, the actual level of intake being determined according to individual characteristics such as sex, age and height. Simple and novel diets can be effective at least temporarily and behavior therapy can make restriction of energy intake easier to achieve.

11.   As present there is no drug treatment for obesity that is clinically safe and useful. Exercise has the same general benefits for obese patients as for all other people, but is not of itself an effective way to loose weight. The surgical treatments of gastric bypass and gastric stapling are methods of enforcing dieting rather than alternatives to it.

12.   When a person succeeds in losing weight, the maintenance of that weight loss must be considered.

13.   In order to prevent obesity, the features which make some individuals more susceptible than others should be addressed, and the environmental factors which predispose to obesity in general must also be tackled. The later requires action and spending at government level; this is warranted because of the economic costs of obesity.

14.   Early effective treatment of moderate obesity in adults is probably the most efficacious strategy for preventing clinically important obesity from developing.

DIET INTERACTION TO THE NERVOUS SYSTEM

1.   Nutrition during fetal and early post natal life is important in determining brain function.

2.   Disturbance of any stage of brain development may cause intellectual and behavioral impairment.

3.   Synthesis of brain protein depends on a constant supply of amino acids in the first two years of life.

4.   Foliate deficiency and genetic factors in the mother cause neural tube defects of the fetus.

5.   Iodine deficiency is the most common profound and avoidable cause of brain damage.

6.   Supply of long chain polyunsaturated fatty acids (PUFAs) especially docosahexaenoic acid (DHA) is crucial to fetal development of brain and visual functions. Deficiency may be linked to later difficulties such a attention deficit disorder dyslexia.

7. Protein energy malnutrition causes stunted growth and poor mental development. Food supplement and mental stimulation in early childhood are beneficial.

8. Iron deficiency anemia impairs cognitive and psychomotor development and should be corrected by iron supplements in childhood.

9. Amino acids are precursors of neurotransmitters and their dietary manipulation may alter mood and behaviour.

10. Disorders of amino acid metabolism lead to brain damage but this may be preventing by maintaining affected babies and children on restricted diets.

11. Adequate supplies of vitamins and minerals are essential for normal brain function.

12. Hydrazine derived drugs may be induce pyridoxine deficiency so that vitamin B_6 supplements are needed, but high doses of these may impair memory and have other side effects.

13. Infants breast fed by strict vegan mothers are at risk of neurological disorders of vitamin B_{12} deficiency. Vegan children should be given vitamin B_{12} supplements. There is risk of vitamin B_{12} deficiency after nitrous oxide anesthesia.

14. Neonatal jaundice should be treated early with riboflavin as bilirubin is toxic to the brain.

15. Favism is a genetic defects infant and breastfeeding mothers with this condition should avoid the faba bean and other hemolytic agents and have a diet rich in antioxidants.

16. Poor diets, deficient in B vitamins, in the elderly and in psychiatric patients may contribute to memory loss, depression, dementia and other common disorders.

17. Vitamin E therapy may prove useful for some brain disorders and may postpone the degenerative diseases aging.

18. Zinc deficiency causes neurophysiologic important. Zinc influences appetite, taste, smell, vision, and mood and may help treat anorexia.

19. Cynogenic components of plants such as poorly detoxified cassava, limit the availability of vitamin B12, cause ataxia and worsen iodine deficiency.

20. A type of pea, lathyrus sativus, eaten in India and Ethiopia, can cause muscle cramps and paraplegia.

CARBOHYDRATES

1. Dietary carbohydrates are classified into four major groups; 1. Free sugars

2. short chain carbohydrates 3. starch and 4. non-starch polysaccharides (NSP).

2. The site, rate and extent of digestion of carbohydrates, which are highly dependent on food processing, have significant implications for health.

3. Starch is subdivided into rapidly digestive starch (RDS) and slowly digestive starch (SDS) and resistant starch (RS).

4. The in-vitro measurement of rapidly available glucose (RAG, RDS and free sugar glucose) provides information about the likely glycaemicresponse to plant foods.

5. The measurement of dietary fiber as NSP provides a good marker of the unrefined plant foods for which benefits to healthcare known.

6. Most of the carbohydrates that reaches the large intestine is fermented by the colonic micro flora, with the production of short chain fatty acids and gases.

PROTEINS

1. Protein is the major functional and structural component of all body cells. It also comprises 10-15% of dietary energy.

2. Proteins are macromolecules consisting of amino acids joined by peptide bonds. Polypeptide chains may consist of as few as two to as many as thousands of amino acid units.

3. During digestion of peptide bonds are hydrolyzed in the small intestine. Eventually free amino acids are secreted in to the blood stream or metabolized in the gut.

4. Intracellular protein synthesis takes place within almost all cells of the body, by the same well characterized mechanism. The mechanisms of protein degradation are more complex and less well characterized. In the adult human protein synthesis and degradation are each more than 250 g daily, compared with a dietary intake of 70 g.

5. The body of the 70 kg. Man contains about 11 kg protein. The protein mass can be influenced by a variety of nutritional and pathological factors. A range of techniques may be used to assess the regulation of protein metabolism.

6. Modulators of protein metabolism include age, growth, dietary intake, injury, disease, hormonal and other factors.

7. The term 'protein requirement' may refer to two different entities; a dilatory allowance or a metabolic need. Requirements for growth and for protein maintaince need to be considered.

8. Amino acids may be categorized as essential, conditionally essential or non essential.

FATS

1. **Roles of dietary fats:** (a) Provision of metabolic energy, 38 kj/gm. (b) Supply of essential nutrients (c) Fat soluble vitamins (d) Fats often improve flavor perception and impart a pleasing texture, thereby improving palatability.

2. Fats in the body play key role in membrane structure and are stored in adipose tissue as a fuel reserve. Fats in foods provide a concentrated form of metabolic energy. They often improve flavor perception and impart a pleasing texture to foods there by increasing the palatability. Dietary fats also supply essential nutrients: fat soluble vitamins and essential fatty acids. The later perform vital body functions but cannot be made in the body and so are essential in the diet.

3. Two primary essential fatty acids linoleic and alpha linolieic acids are converted in the tissues in the longer chain length, more highly unsaturated fatty acids that are located principally in biological membranes. These contribute to membrane structure and are also converted in to oxygenated fatty acids. These hormones like substances are locally produced and have many functions in cellular communication. They are involved in the regulation of blood coagulation, blood pressure, muscular contraction, immune function and inflammatory responses and diverse aspects of the regulation of cellular metabolism.

4. Fats are normally efficiently digested and absorbed; when normal fat absorption is impaired, fats containing short and medium chain length fatty acids can be effectively utilized. The products of fat digestion are resynthesized into lipids that are transported in blood stream as lipoproteins. The protein components of lipoprotein interact with specific receptors on utilization of the products of fat digestion. Malfunction of these receptors as a result of gene defects, dietary imbalance or ineffective energy expenditure results in aberrations in lipoprotein metabolism that can lead to chronic diseases. Overconsumption of fats without concomitant increases in the utilization of the energy leads to excessive storage in adipose tissue and finally obesity. Tailor-made fats with zero or reduced energy value are now being introduced in to foods.

5. Different foods supply different amounts of fat with widely differing fatty acid composition. Improvement in food compositions databases and better information on food labels now allow selection of foods to help consumers regulate the amount of fat eaten and ensure an appropriate balance of fatty acids.

ALCOHOL

1. Alcohol consumption is a feature of many societies, with wide variations between the later in drinking habits. Actual intake is difficult to estimate.

2. Rates of absorption and metabolism vary widely between individuals. The differences are partly genetic in origin, but also depend on gender, previous level of alcohol intake and other factors.

3. Alcohol is mainly metabolized in the liver and four metabolic routes have been described.

4. In heavy drinkers the induced microsomal ethanol oxidizing system (MEOS) plays a major part in ethanol metabolism and also considerable effects and drug metabolism.

5. The tolerance of alcohol varies between ethnic groups. There is also a familial tendency to alcoholism, but environmental factors are nevertheless important.

6. Chronic heavy alcohol intake has a variety of toxic effects on body systems and on the developing fetus in pregnant women.

7. Alcohol abuse has costly social and economic consequences.

8. Carbohydrate, lipids, protein and sex hormone metabolism are affected by alcohol.

9. Alcohol is associated with an increased risk of chronic disease in adults. However, moderate alcohol consumption seems to have a protective effect with regard to coronary heart disease.

ENERGY BALANCE

1. Energy balance is the difference between metabolizable energy intake and total energy expenditure. It is strongly related to macronutrient balances, and the sum of the individual substrate balances expressed as energy, must be equivalent to the overall energy balance.

2. Modern lifestyles have led to considerable changes in what food is eaten and circumstances in which it is consumed, contributing to poor control of food takes.

3. Energy expenditure is less effective than food intake as a control mechanism for energy balance.

4. Although the physiological mechanisms for controlling intake, i.e., hunger and satiety exists in man, appetite as a powerful and poorly controlled stimulates to eat. Satiety itself is to some extent a conditional reflex.

5. Short term regulation of energy balance is poor, but (in most people) long term regulation is accurate. The mechanism is unknown, but must

include conscious alterations in lifestyle to correct unwanted changes in body weight.

6. During long periods of energy imbalance the weight gained or lost is initially glycogen plus water with an energy density of ~1.0 kcal/g if the imbalance continues, after a week the tissue gained or lost is a mixture of fat, water and protein with an energy density of ~7 kcal/g.

7. A high fat (energy dense) diet promotes weight gain, because it promotes increased energy intake. However, a low fat diet consumed in excess also causes an increase in body fat, because far oxidation is inhibited.

8. Under nutrition leads to a decrease in energy expenditure. Part of the decrease in metabolic rate is related to weight loss.

9. In over feeding, although some of the excess energy intake will be stored in adipose tissues, there are compensatory increases in energy expenditure.

WATER AND MONOVALENT ELECTROLYTES

1. Water is the main constituent of the body of the 45 liters. Of water in an average 70 kg. Man, about 30 liters are intercellular fluid (ICF) and about 50 liters are extracellular fluid (ECF). The electrolyte ions of sodium, potassium and chloride are responsible for most of the osmolality of body fluids and their distribution determines the volumes of ICF and ECF.

2. The total amount of water in the body is ultimately determined by the quantities of Na^+ and K^+ and their accompanying anions. The mechanisms for controlling these amounts are oral intake and urine excretion to a lesser extent faces and losses from skin and lungs.

3. Correction by the body of abnormal Na^+, Cl^- and K^+ content takes much longer than correction of water content. Common salt comprises most of the Na^+ and Cl^- intake; K^+ is ingested as a constituent of nearly all forms of natural food. The kidney is mainly responsible for excretion of these electrolytes.

4. Possible deficit states include water depletion (pure water loss) saline deficiency (isotonic loss of Na^+, Cl^- and water, traditionally and misleadingly referred to as dehydration) chloride depletion and potassium depletion. For water depletion treatment depends on the cause; oral electrolyte supplements may be used in the later three conditions.

5. An excess of body water alone is rare. Treatment is severe cases is aimed at increasing plasma Na^+ and osmolality to normal levels. Saline excess is the commonest abnormality of water and electrolyte metabolism conditions. Potassium excess is rare than K^+ depletion.

6. There are important differences between water and electrolyte values in infants and adults. Saline depletion is common in infants, but they are also at greater risk than adults of developing water depletion and saline excess.

BONE MINERAL

1. The skeleton is not inert. There is one organic component to bone. Formation and resorption of bone occurs continuously by means of cells influenced by mechanical, nutritional and hormonal factors. The skeleton has a biochemical as well as mechanical function.

2. The important bone minerals are calcium, phosphorus and magnesium. Mineralization can occur in two apparently separate processes. In the first, crystals are formed within vesicles and in the second the mineral is laid down an ordered way on the organized fibrils of the organic collagen matrix. The later also contains non collagen substances including the bone morphogenetic proteins.

3. Bone turnover is controlled by osteoblasts, which are responsible for bone formation and respond to endocrine, nutritional and mechanical signals by osteoclasts which are involved in resorpration and by osteocytes.

4. Calcium is the most important mineral in the skeleton. Calcium balance is determined by exchanges between the skeleton. The intestine and the kidney; these fluxes are controlled by 1, 25-Dihydroxycholercalciferal, Parathyroid hormone and calcitonin.

5. The skeleton is also affected by hormones such as growth hormone, sex hormones, thyroxine, corticosteroid and insulin. Locally acting cytokines also produce changes in bone metabolism. Calcium homeostasis is distributed in different ways by various disease states.

6. Up to the time peak bone mass (possibly at about 30 years of age) new bone formation exceeds resorption. Determinants of peak bone mass include genetic inheritance, nutrition and exercise; it is decreased by immobility, smoking and alcoholism.

7. Loss of bones occurs in all subjects after the age of peak bone mass, but it is accelerated in menopausal and post menopausal women because of estrogen deficiency. Factors which increase bone loss are largely the same as those which decrease peak bone mass.

8. In osteoporoses the amount of bone per unit volume decreases but the composition remains same. There is a predisposition to fracture and the amount of bone is reduced to more than 2.5 standard deviations below the mean for young adults.

9. Phosphorus a necessarily constituent of many biochemical processes is widely distributed throughout the body but 85% is found in the skeleton.

10. The most important factor in phosphate balance is re absorption by the renal tubules. Causes of abnormal phosphate balance may be renal or dietary and inherited or acquired.

11. Magnesium is distributed through the soft tissues but 60% is contained in the skeleton.

12. The plasma magnesium level is remarkably constant and depends upon rapid adaption by the kidneys. Hypomagnesaemia occurs in association with hypocalcaemia but also may be found in its own. The commonest cause is probably malabsorption. Hypermagnaesaemia is most commonly seen in patients with renal glomerular failure.

13. Other minerals normally present in bone, which affect skeletal health, include copper, aluminium and fluorine. Strontium is of interest because its radioactive form, strontium-90 is produced in nuclear explosions and is incorporated in to the skeleton.

IRON, ZINC AND OTHER TRACE ELEMENTS

1. Iron absorption is carefully regulated. In iron deficiency absorption can increase up to a certain level. When iron stores reach an upper physiological level no more iron is absorbed, regardless of dietary iron content and bioavailability.

2. Menstrual losses vary considerably between different women but are very constant in an individual woman, making it difficult for her to judge weather losses are heavy, normal or small.

3. Iron requirements are especially high in infants of 6-24 months of age, in adolescents and adult menstruating and pregnant women.

4. The composition of the diet has a decisive influence on iron absorption.

5. Absence of iron stores is associated with an iron deficient erythropoiesis and a compromised supply of iron, not only to red cell precursors but also to other tissues.

6. Iron deficiency can be established by measurement of a low serum ferritin value, absence of stainable bone marrow iron or a response to therapeutic iron.

7. Worldwide the main cause of iron deficiency is nutritional; dietary iron absorption cannot cover physiological iron requirements.

8. In clinical practice a cause for iron deficiency should always be sought.

9. A mild iron deficiency without anemia is associated with reduced physical performance and changes in mood and the ability to concentrate.

10. Iron requirements in pregnancy can be covered from a good diet if iron stores contain about 500 mg. Since most women have only small or negligible stores, even an excellent diet cannot provide all the iron needed by mother and fetus. Regular iron supplement in pregnancy is therefore needed by most women.

11. It is irrational to treat already existing iron deficiency anemia by diet alone. Iron supplementation is required using a well absorbed iron preparation at an adequate dosage.

12. Several factors need to be considered in the choice of strategies to prevent iron deficiency. Strategies include food education, iron fortification and iron supplementation.

13. Hereditary haemochromatosis is not a nutritional problem but an important clinical problem and action should aim at early detection and treatment.

14. Zinc is essential for a vast number of biochemical processes and plays a central role in growth and development.

15. The body possesses effective mechanisms for maintaining zinc homeostasis over a large number of intakes.

16. However, the utilization of dietary zinc is highly dependent on diet composition with risk of inadequate supply of zinc where there is a high intake of unrefined cereals with high phytate content and a low intake of animal protein.

17. The risk of inadequate zinc supply is highest during periods of rapid growth such as infancy, pubertal growth and pregnancy.

18. High intake of zinc in the form of supplements can interfere with absorption and utilization of other trace elements.

19. Copper is widely distributed in the body and plays a role in some fundamental functions.

20. Cu deficiency is seen most frequently in preterm infants and in appropriately fed term infants. Menkes syndrome arises from a genetic defect in Cu transport and metabolism.

21. Which chronic overexposure, Cu accumulates in the liver leading to hepatic necrosis or cirrhosis with liver failure. Wilson's disease arises from a genetic defect in Cu metabolism, with reduced biliary accretion of Cu and accumulation in the liver, eyes, brain and kidneys.

22. Selenium is essential for the formation of 12-30 mammalian proteins. The total body content of Se is 3-30 mg and varies according to geochemical environment and dietary intake.

23. Deficiency in Se intake is associated with cardiomyopathic disease. There is substantial evidence that Se supplementation can reduce the incidence of some cancers.

24. The average human body contains 10-20 mg of manganese; 25% of this is in bone, the rest being distributed throughout the tissues. It is found particularly in vegetable based diets and in tea. There is no well documented case of human Mn deficiency, but excess Mn is linked with psychosis and with parkinsonian symptoms.

25. Molybdenum occurs as a co-factor in three major human and animal enzymes. The impact of Mo on human health is associated with genetic errors of metabolism rather than deficiency or toxicity.

26. A low intake of fluoride is linked with dental caries, while chronic excessive intake causes bone disease and joint abnormalities. The intake range of fluoride that benefits the teeth is narrow, and recommendations regarding fluoride intake from sources other than water have been amended to take account of local levels of water fluoride.

27. It is thought that chromium might potentiate the action of insulin. Cr deficiency has been described. Trivalent Cr has a low toxicity, but hexavalent Cr is more toxic and occupational exposure is an industrial problem.

VITAMINS

1. The vitamin A family consists of a series of derivations of all-trans-retional.

2. Hundreds of carotenoids exist in nature but only about 50 of them can be converted to vitamin A; the best known is B-carotene.

3. Performed vitamin A is absorbed along with fat in the diet, at an efficiency of more than 80%.

4. Bioavailability of carotenoids varies from 5-50%.

5. Vitamin A circulates in the plasma mainly as all transretinol attached to retional-binding protein (RBP).

6. Caretinoids circulate in LDL and HDL lipoproteins.

7. Vitamin A in the body 80-905 stored in the liver in parenchymal cells and special stellate cells.

8. β-carotene is converted to vitamin A by oxidative enzymatic cleavage.

9. Vitamin A undergoes many reversible transformation and isomerizations. Most forms have specific RBPs.

10. The main function of vitamin A is in vision where it forms the light sensitive complex rhodopsin in the eye.

11. Other functions are in cells differentiation, embryogenesis and the immune response.

12. Dietary sources of vitamin A are liver, dairy products and oily fish.

13. Dietary sources of carotenoids are yellow/orange vegetables and fruit and dark green leafy vegetables.

14. Signs of inadequacy are increased rate of infections and mortality.

15. Signs of deficiency are night blindness, xeropthalmia and bilots spots and the conjunctiva of the eye.

16. Treatment of deficiency is by oral doses of retinyl palmitate in oil.

17. Toxicity occurs at 20 times the RDA in children and 100 times the RDA in adults, but recovery usually occurs slowly when intake ceases.

18. High doses of vitamin A in pregnancy are teratogenic causing fetal resorption, abortion, birth defects and severe learning disability.

19. Cholecaciferol, Vitamin D_3, is the natural form of vitamin D synthesized by the action of sunlight on the skin.

20. Vitamin D controls calcium metabolism by stimulating absorption in the gut and resorption from bone and by allowing calcium to circulate, increases bone formation.

21. Calcium is important in nerve and muscle function, so low calcium intake may result in bone resorption.

22. Vitamin D deficiency results in poorly mineralized skeleton, causing rickets in children and osteomalacia in adults.

23. Vitamin D contributes to the regulation of the formation of immune cells in the blood.

24. Plasma 25 (OH) D_3 is an index of availability of vitamin D. The normal range is 8-60 ng/ml. Levels below 10 ng/ml indicate impending deficiency; toxicity occurs at levels above 150 ng/ml.

25. Natural food sources of vitamin D egg yolk, oily fish, butter and milk but these are relatively trivial.

26. The major food source is supplemented margarines and spreads.

27. Sunlight is the best source of vitamin D.

28. Supplements are recommended for young children, pregnant woman, the elderly and anyone who is housebound or gets no sunlight.

29. An introduction to free radicals and antioxidants mechanisms is given for vitamin E.

30. The major dietary sources of vitamin E are vegetable oils.

31. There are eight naturally occurring compounds: four tocopherol, so this is the most important dietary form.

32. Binding proteins preferentially select alpha tocopherol, so this is the most important dietary form.

33. A major function of vitamin A is as an antioxidant preventing PUFA lipid peroxidation.

34. Other functions include maintaining cell membrane integrity, involvement in anti inflammatory and immune systems, and in DNA synthesis and cell signaling.

35. Vitamin E deficiency causes damage to cell membranes and leakage of cell contents, resulting in myopathies, neuropathis and liver necrosis.

36. Suboptimal levels of vitamin E may be linked with atherogenesis. Coronary heart disease some cancers and other degenerative and neurological disorders.

37. Naturally occurring vitamin K compounds possesses a napthoquinone ring with different side chains.

38. Phylloquinone is synthesized by plants and has phytyl side chain.

39. Menaquinines are synthesized by bacteria and have prenyl units in a side chain.

40. Dietary sources of phylloqunone include vegetables, especially dark green leafy vegetables (DGLV) and some vegetable oils.

41. Menaquinones are found infermanted food such as cheese and yogurt and in ruminant liver.

42. Menaquinones are synthesized by bacteria in the colon, but their bioavailability is uncertain.

43. Absorption in the proximal intestine is 80% efficient for free phyloquinone but much less from DGLV where it is tightly bound in chloroplasts.

44. Vitamin K is absorbed as chylomicrons through the lymph to the blood where it is carried to the liver and other tissues.

45. The liver has limited storage for phylloquinone (10%); menaquonine stores comprise 90% but their functional significance remains uncertain.

46. 60-70% of phylloquinone in excreted in urine and faces within a few days, suggesting the need for constant replacement.

47. Vitamin K is a cofactor for the conversion of Glu to Gla in the synthesis of calcium binding proteins; the coagulation factors II (prothrombin) VII, IX, and X feedback proteins C and S and protein Z of uncertain function.

48. Vitamin K is involved in the synthesis of Gla protein in osteocalcin by osteoblasts, but its importance in bone health is not yet understood.

49. Other Gla proteins protect soft tissues from calcification and control fetal bone development.

50. Vitamin K deficiency is responsible for haemorrhagic disease of the newborn, also known as a vitamin K deficiency bleeding (VKDB). Routine prophylaxix is advised in most countries.

51. Suboptimal vitamin K status may be linked to bone disease.

52. The major sources of vitamin C (ascorbic acid) are fresh vegetables and fruit. Although potatoes do not contain large amounts, they can be an important source where they are a staple food.

53. Ascorbic acid is absorbed by a sodium dependant mechanism. At high intakes saturation occurs and excretion in the urine rises.

54. Ascorbic acid can be crystallized from D-glucose and D-galactose. The naturally occurring vitamin is L-xylo-ascorbic acid. Ascorbic acid serves as an antioxidant and as a cofactor in hydroxylation reactions.

55. In man, the highest vitamin C concentrates are found in the adrenal and pituitary glands but there are substantial amounts in the eyes, kidneys, liver, and spleen and in other parts of the brain.

56. Generally plasma ascorbic acid levels reflect daily intake and leukocyte levels reflect tissue stores. These values usually correlate in population studies, so measurement of plasma levels is usually used to determine vitamin C status as it is less labour intensive.

57. Confirmed scurvy is usually associated with ascorbic acid levels in plasma more than 7 micromole/lit and levels in leucocytes of around zero. Other diseases smoking and surgical stress all reduce plasma and leukocytes ascorbic acid levels independently of vitamin C status.

58. It seems likely that any particular plasma concentration or intake of vitamin C is optimal for all situations.

59. It is still not known to what extent vitamin C has a physiological role in the immune response. Studies investigating whether vitamin C supplements affect the common cold have shown very varied results.

60. In the UK the recommended reference nutrient intake (RNI) is 40 mg/day for adults, compared with a US recommended dietary allowance (RDA) of 60 mg/day. These recommendations are adjusted for pregnancy and lactation for infants and children and for smokers.

61. Thiamin is present in all natural foods. It is mostly obtained from unrefined cereal grains or starchy roots and tubers by people in

developing countries, and from fortified cereals and breads by people consuming western diets.

62. Thiamin is absorbed from the upper intestine. There are two mechanisms: an active saturable, sodium dependant process, and a passive absorption which operates at oral intakes of more than 5 mg/day. This has implications for alcoholic patients, as ethanol inhibits the active but not the passive process.

63. Of the thiamin in the body, 90% is present as thiamin diphosphate (TDP) and approximately 105 is present in the nervous tissues as thiamin triphosphate (TTP).

64. TDP is especially important for carbohydrate metabolism, and thiamin seems to have specific function in neural tissues.

65. The erythrocyte transketolase assay is the most widely used functional measurement of thiamin status. This assay becomes more sensitive as deficiency becomes more severe.

66. Thiamin deficiency may lead to wet (cardiac) or dry (neurological) beriberi; however, both cardiac and neurological functions are disturbed. Wet beriberi responds rapidly to thiamin therapy. In dry beriberi the response is slow or absent, long term treatment may be required and cure may not be complete.

67. Thiamin deficiency is associated with alcoholism. Malabsorption of the vitamin and excessive carbohydrate intake can precipitate the wernicke-korsakoff syndrome. Reversibility of the condition by means of thiamin therapy depends on the stage of the disease.

68. Because of the essential role of thiamin in carbohydrate metabolism, recommended intakes are expressed in terms of energy intake (mg/4.2 mj or mg/1000 kcal). Where values have been calculated for western populations, the thiamin requirements may need to be adjusted upwards for people in developing countries to take into account relatively higher carbohydrate consumption.

69. Riboflavin is principally found in dairy products. Cereals which have not been fortified are poor in riboflavin, but fortified cereals may be the main source of the vitamin in developed countries.

70. Riboflavin is absorbed in the proximal intestine by a sodium dependant saturable process. Flavin mononucleotide (FMN) and flavin adenine dinuclotide (FAD) are involved in oxidation reduction reactions in a number of metabolic pathways.

71. Deficiency of riboflavin leads to ariboflavinosis characterized by the retarded growth and often inflammation of the epithelium at the border between skin and mucous membranes.

72. Riboflavin status is most commonly assessed by measurement of the activity coefficient of erythrocyte glutathione reductase. Requirements may increase with physical activity and negative nitrogen balance.

73. Niacin is obtained in the diet both as the vitamin, nicotinic acid and the provitamin, tryptophan thus dietary niacin is quantified as niacin equivalents (NE). Cereals such as oatmeal, rice and wheat provide moderate amounts. Treatment of maize with lime converts poorly bioavailable niacytin to niacin.

74. Nicotinic acid is absorbed from the stomach and both nicotinic acid and nicotinamide are absorbed from the small intestine by a sodium dependant process.

75. Niacin status is best assessed by determination of the molar ratio in urine of the metabolites N-methylnicotinamide (NMN) and N- methyl-2 pyridone-5-carboximide (2-pyrodone).

76. Deficiency of niacin causes pellagra and classic pellagra responds dramatically to nicotinamide or nicotinic acid.

77. Excess nicotinic acid causes vasodilation of blood vessels and a transient fall in blood pressure.

78. Vitamin B_6 plays an important part in many reactions an amino acid metabolism, and in glycogen metabolism. It also has a role in the activity of nuclear acting hormones.

79. Malabsorption, celiac disease, renal dialysis chronic alcoholism are risk factors for vitamin B_6 deficiency along with deficiency of other water soluble vitamins. Biochemical signs of B_6 deficiency or inadequacy have been found in 10-20% of the healthy population. The plasma concentration of vitamin B_6 decreases with age, and several drugs are associated with B_6 depletion.

80. While several months are available for assessing vitamin B_6 status, the most useful indices in practice are the plasma concentration of pyridoxal phosphate and the erythrocyte transaminase activation coefficient.

81. In adults the minimum replacement requirement is about 350 microgram/day (for a 70 kg man) and the reference nutrient intake (RNI) based on 15-16 microgram of protein is 1.5-1.6 mg/day. There is a need for further research to determine infant requirements; human milk which is presumably adequate actually provides much less than calculated requirements.

82. Doses of 50-200 mg/day are antiemetic and used to treat nausea associated with radiotherapy and pregnancy. However there is little evidence for the benefit of vit. B_6 therapy in pregnancy sickness, or for its effectiveness in carpal tunnel and premenstrual syndrome. High doses of B_6 are associated with neurotoxicity.

83. Rich sources include meat, fish potatoes, bananas and pulses.

84. Pernicious anemia of pregnancy was first recognized in the 1930s and more than 50 years later the cause was established as being an increase in the catabolism of folate during pregnancy.

85. Folate plays an essential role in providing methylene groups for pyrimidine synthesis, formyl groups for purine synthesis as part of the DNA cycle and methyl groups for the methylation cycle.

86. It is usually presented to cells in the 5-methyltetrahydrofolate form and this is usually the only form of folate in the plasma.

87. Formerly it was believed that the main adverse outcome of a folate deficiency was megaloblastic anemia. It is now evident that even moderate reductions in folate status in pregnancy, to levels hither to consider being within the normal range, increase the risk of fetal neural tube defects. A link between moderate reduction in folate status and risk of cardiovascular disease has also been established.

88. Low folate status is associated with poor dietary intake impaired absorption (e.g., in celiac disease and tropical sprue) pregnancy treatment with some drugs especially anticonvulsants and alcohol abuse.

89. Folate status is assessed by measuring the level of plasma or serum. However the red cell folate level is the better index.

90. There is no particularly good source of folate in the diet (except liver). Synthetic folic acid used in supplements and fortified food is thought to be 100% available compared with 50% bioavailability of natural folate. Notably, folic acid was used in the intervention trials where there was prevention of NTDs.

91. Current recommendations for intake will almost certainly be revised upwards. With regard to the prevention of NTDs supplements of 400 microgram/day or 4.0 mg/day where there is a history of NTD are recommended these must be taken periconceptionally.

92. Suggestions that the diet should be fortified with folic acid raise the question of possible toxicity. High amounts in appropriately treat the anemia of vitamin B_{12} deficiency so that the condition goes undiagnosed, they have been implicated in reducing the effectiveness of anticonvulsants used to stabilize epilepsy and they might reduce the effectiveness of anifolates.

93. Vitamin B_{12} (cobalamin) participates in enzymatic reactions in two forms. Lack of one of the vitamin B_{12} dependant enzymes methylmalonyl CoA mutase is associates with clinical problems and mental retardation. The other vitamin B_{12} dependant enzyme, methionine synthase, is a constituent of the methylation cycle.

94. The capacity of vitamin B_{12} absorption is limited by the specific B_{12} receptors in the ileum. Transcobalamin II (TC II) transports the vitamin to the liver for storage and various tissues where it is utilized.

95. Very little excretion in the urine occurs. There is a small but important excretion in bile of vitamin B_{12} which is reabsorbed.

96. Severe vitamin B_{12} deficiency leads to megaloblastic anemia or sub-acute combined degeneration (SCD). It is believed by most workers in the field that absence of methionine synthetase is involved in both conditions.

97. Decreased vitamin B_{12} status may be caused by inadequate diet, intestinal malabsorption of various forms and interaction with some drugs and with alcohol.

98. Radiometric assay of vitamin B_{12} is now widely Measurement of the plasma levels of B_{12} and folate is particularly important for the differential diagnosis of the anemia's associated with their deficiencies.

99. Vitamin B_{12} only appears in meats or foods of animal origin such as dairy produce. The current recommended dietary allowances in most countries are between 1 and 2 microgram per day.

100. Vegetarians but especially vegans should take an oral supplement of vitamin B_{12}. Patients with mal absorption syndrome is usually treated with an intramuscular injection of 100-1000 microgram every month or second month.

101. The imidazole moiety of biotin is important as the binding site to the egg white protein avidin. Biotin is involved in fatty acid synthesis and metabolism gluconeogenesis and branched chain amino acid metabolism.

102. Human biotin deficiency is rare and leads amongst other symptoms to a dry scaly dermatitis. Biotin levels can be measured in whole blood or urine by microbiological assay.

103. Dietary sources include liver, egg yolk, soy flower, cereals and yeast. Bioavailability is partly determined by binders in food.

104. As a constituent of CoA and its esters, pantothenic acid is essential for several reactions involved in lipid and carbohydrate metabolism.

105. Human pantothenic acid deficiency is very rare, but signs include par aesthesia with 'burning feet'. Pantothenic acid levels in blood or urine are measured by microbiological assay or radioimmuno assay.

106. Pantothenic acid is found widely, but especially in animal products, whole grains and legumes.

NUTRITION AND GENETICS (HUMAN GENETICS)

1. Increasing knowledge about the genetics basis for variations in nutrient requirements and responses to diet in transforming concepts of nutrition and disease.

2. The interactions between nutrients and genes may be categorized in to six main types, relating to; (i) intrinsic universal features of the human genotype (ii) Intrinsic individual genetic differences (iii) Fetal genetic programming caused by the maternal diet (iv) Intrinsically low levels of genetic expression, (v) influence of the diet on genetic expression and, (vi) Dietary influences on the genetic changes which occur with ageing.

3. Human genes lack the coding for synthesizing a variety of essential and conditionally essential nutrients which must therefore be provided in the diet.

4. There are notable racial differences for instance with regard to lactase deficiency and alcohol metabolism. The alteration of a single base can have far reaching effects.

5. Some important genetic features are very common and others are very rare. Genes which lead to minor variations in nutrient requirements may have a high incidence for instance about 10 % of people have relatively higher folate needs because of a valine substitution for alanine in the enzymes methyltetrahydrofolate reductase.

6. Macronutrients, especially the polyunsaturated fatty acids, regulate the expression of the genes for enzymes, promoters and inhibitors involved in metabolic pathways.

7. Oxidative damage to DNA can be assayed by variety of methods, which are now being used to investigate the protective effects of specific dietary components against such damage.

8. There is now increased understanding of the genetic defects involved in colorectal cancer. Individuals at high risk for this disease can now identified and offered regular screening and dietary pharmaceutical interventions.

9. Genetic inheritance contributes to 25-40% of the variation in body fat and has an even greater effect on the pattern of fat distribution in the body. At present more than 40 genes are being investigated in studies of human obesity, but the protein leptin has a significant role.

FUELS OF THE TISSUES

1. The energy cost of metabolic cycles can be estimated. The increased activity of these cycles' accounts for a substantial part of the rise in basal metabolic rate is severally injured patients.

2. The metabolic function of a tissue is affected by a supply of substrates and hormones from a variety of other tissues.

3. The type of fuel utilized by a tissue depends partly on availability but some tissues have specific fuel requirements or fuel preferences which may be hierarchical.

4. At tissue level metabolism may be regulated primarily by blood flow, substrate supply or competition between circulating fuels.

5. The interogan flux of some substrates is far greater than the dietary intake. Thus the tissues producing these substrates are likely to be of greater importance than the dietary intake in controlling the supply.

6. For some tissues an extravascular supply of fuels is important.

7. Changes in fuel requirements for a specific tissue may have major effects on the whole metabolism.

8. Metabolism is also regulated by circulating and local hormones and cytokines.

9. All tissues receive and send signals to other tissues some signals have general effects and some have specific targets.

ENERGY INTAKE AND EXPENDITURE

1. Energy in foods is provided by carbohydrates, proteins, fats and alcohol. Of the chemical energy in foods, 5-10% is lost through the faces and urine.

2. The energy available to the body or metabolizable energy is on average 16 kj/g of carbohydrate, 17 kj/g of protein, 37 kj/g of fat and 29 kj/g of alcohol. These figures very slightly according to the type of carbohydrate, protein or fat in the diet.

3. The energy used in the body or energy expenditure, is usually measured by indirect calorimetry, which involves measuring the oxygen used and or carbon dioxide produced by an individual in the laboratory.

4. The doubly labeled water method is a new technique which allows energy expenditure to be recorded over periods of 2-3 weeks while subjects follow their normal activities. It can be used in young children, athletes, pregnant women and hospital patients.

5. Energy expenditure is primarily determined body size, body composition and physical activity. Hormones drugs and environmental temperature may also be important in some conditions.

6. Energy requirements increase by around 800 kj/day in the last trimester of pregnancy, and by around 2100 kj/day during full lactation.

COMPOSITION OF THE BODY

1. During both intrauterine and postnatal growth the body changes in composition as well as in size. Normal growth and chemical maturation may be disturbed by either malnutrition or disease.

2. Information about the chemical composition of the adult human body is based on a small number of cadaver analyses. The fat free body has a fairly constant chemical composition. Laboratory techniques can able the estimation of fat free mass from the measured density, water or potassium content of the body or ideally all three of these. When extra fat is stored a quarter of the weight gained is nonfat tissues.

3. For clinical purposes adiposity is conventially estimated using Quetelets Index, or BMI, i.e., W/H_2, or by measurement of subcutaneous skin folds. New instruments measure electrical conductivity of the body which indicates the water content of the body and hence fat free mass.

4. Imaging techniques, using X-rays or magnetic resonance can provide quantitative estimates of the amount of fat, bone and non-skeletal fat free tissue, as well as information about the spatial distribution of these tissues.

FOOD COMPOSITION TABLES AND NUTRITIONAL DATABASES

1. Food composition tables and increasingly nutritional databases are essential resources for work in nutrition.

2. Data bases may be compiled using direct or indirect methods or a combination of the two. Whatever method is used, criteria must be applied with regard to sampling and analysis.

3. There are differences in organization between printed nutritional tables and computerized databases.

4. There are three major ways in which the compilations are used: for nutritional analysis, nutritional synthesis and as a primary source for educational or regulatory purposes.

5. The main limitations of food composition data relate to the variability in the composition of foods and the incomplete coverage of all the foods that make up the human diet.

6. The growth of nutritional epidemiology has led to increased international use of databases. Consequently, attention has been paid to tissues concerning the unequivocal identification of foods, compatibility of data, data interchange and data quality.

7.   The user of any compilation of food composition data needs to be aware of the conventions of the database of the criteria used in its composition and of the limitations in the accuracy.

METHODS AND VALIDITY OF DIETARY ASSESSMENT

1.   Before a dietary assessment is undertaken, it is necessary to consider the underlying purpose of the assessment, what is to be measured, in whom, over what time period, and how the data are to be collected.

2.   There are five levels at which food availability and consumption can be conveniently assessed. Levels I and II are the national levels of domestic food production and total food available, levels III and IV are the household levels of food purchases and food available in the household and level V is that of individual food consumption.

3.   Data at national levels include information on food production and are more usefully on the total food available.

4.   At household level four main techniques of the data collection are used; food accounts, inventories, household record and list recall.

5.   Techniques for individual dietary assessments may be conveniently classified as prospective or retrospective and each method has strengths and limitations.

6.   There has been a growing recognition that all measures of diet have sources of bias. Attempts are made to identify errors and adjust for them, in order to assess more accurately the relationships between diet and health.

7.   When estimates of dietary intake are reported it is important that a full description is given of the method of assessment.

CEREALS AND CEREAL PRODUCTS

1.   Cereals remain the staple food in most diets, even given their decline in importance as a country because more affluent, where cereals are produced and to a large extent their importance depends on climate. On the basis of overall production, wheat is the major cereal, closely followed by rice and maize.

2.   Cereals are seeds, with a typical anatomy although there are important structural differences between the different grains.

3.   Cereals undergo a variety of milling, hulling and pearling processes before human consumption. Milling does not have major effects on the nutrients within the grains. However, the fractionation process, which may follow milling, separates the components of the grain and is

significant nutritionally because the nutrients are not distributed evenly throughout the seed.

4. The composition of cereal products is affected by the processes used in their production, some of which are very vigorous and also by cooking method.

5. In the UK cereals provide about 30% of energy, 25% of protein, about 50% of available carbohydrates and about 10% of fat. They make a major contribution to dietary fiber as non-starch polysaccharides. The nutritional role of flour in the UK is enhanced because of fortification with vitamins, calcium and iron.

MEAT, FISH, EGGS AND NOVEL PROTEINS

1. Meat has formed an important part of the human diet throughout evolution, and still has considerable cultural significance.

2. In Europe and the USA the major carcass meats are cattle, sheep and pigs. In the Middle East, Africa and the Indian subcontinent, goats, camels and water buffalo are more important. Worldwide, a large number of species of wild animals and game are eaten. In the UK and other countries meat from poultry has become increasingly important with the development of large scale intensive and economical production regimes.

3. As the nutrients in meat are present in different concentrations in the fat and the lean the composition depends on the fat to lean ratio. Meats are usually regarded as protein foods. They are important sources of fat and vitamin B_{12}. The high bioavailability of their inorganic nutrients is a significant feature of meats.

4. Fish may be categorized for nutritional purposes as white fatty or cartilaginous fish. Within these groups there is considerable similarity of nutritional composition. Fish are regarded as a source of good quality protein. Consumption of fish in the UK is relatively low, although it has risen slightly in recent years.

5. In the UK most hens' eggs are produced under battery conditions. The composition of eggs produced under less intensive systems shows minor differences from that of battery eggs, but none that are nutritionally significant.

6. Eggs are a food of high biological value and can be used to improve the supply of animal protein. They are extensively used in processed foods.

7. Novel protein foods have been developed from isolates of plant proteins and by microbiological methods.

MILK AND MILK PRODUCTS, FATS AND OILS

1. Milk from many species has an important role in the human diet for thousands of years.

2. Liquid milk is unstable so many of the milk products which evolved are fermented and could be stored and transported.

3. In the UK milk production is mainly from cows, usually Friesians, with a small but growing contribution from goats and sheep. The nutritional safety of milk is still heavily dependent on the maintains of hygienic standards in production.

4. There is now a substantial demand for skimmed and semi-skimmed milk.

5. There are number of processes which reduce the water content of milk and make it more stable, producing dried, condensed evaporated and ultra heat treated (UHT) milk.

6. Different types of cream vary with respect to their fat content and also according to production methods which modify the physical properties.

7. Cheese is produced by coagulation of milk proteins to form a curd. Because of the many variations in the cheese production, the composition of chesses varies considerably, even sometimes between two samples of the same named cheese.

8. Milk may be fermented with lactobacilli, to produce yoghurt for example.

9. Products such as butter, based on the components of milk are important in the diet both as foods and ingredients.

10. Milk is the sole food for almost all young mammals and the milk of each species contains all the ingredients needed for growth in that species. Major proteins include casein, lactalbumin and a range of immunoglobulin. In man the latter are believed to be important for neonatal immunity. The fat content in milk varies between species and in ruminants contains only low levels of unsaturated fatty acids. Human milk has lower levels of inorganic nutrients than cow's milk, especially of phosphorus. Milk contains both fat soluble and water soluble vitamins; the levels of different vitamins are altered by processing and storage.

11. Milk and its products are significant sources of nutrients and it is important, especially with regard to children, that this role is not undermined by inappropriate interpretation of nutritional guidelines for fat intake.

12. Lipids isolated from animal products tend to be solid fats, such as butter, lard, ghee and suet, and those from plants are usually oils, but the distinction lies in their fatty acid and triacylglycerol composition rather than their origin.

13. Some plant oils are derived from the flesh of the fruit such as olive oil

and some from the seeds.

14. In recent years there has been increased consumption of margarine made from highly unsaturated oils, such as maize and sunflower, because of the beneficial effects upon serum cholesterol.

15. The main characteristic of nutritional importance in the composition of fats is the fatty acid content.

16. The concentration of the fat in a food or diet is the principal determination of its energy density. Fats provide the essential fatty acids and contribute to the absorption of fat soluble vitamins. There is an important association between fat in foods and palatability, which has implications for attempts to lower dietary fat intake.

PREGNANCY AND LACTATION

1. Recommended weight gain in pregnancy depends on pre-pregnancy weight and may range from 7 kg in the overweight to 18 kg in the underweight.

2. The energy cost of pregnancy in different environments may vary fourfold from 78 MJ to 286 MJ.

3. WHO recommends an extra intake of 1.3 MJ daily throughout pregnancy?

4. The adipose tissue stored during pregnancy is normally sufficient for fetal growth and 6 months of lactation.

5. In developed countries, each pregnancy leads to an average post pregnancy weight gain of 1 kg.

6. Breast feeding (in developed countries) helps prevent this progressive weight gain if extra energy intake is limited to 1.3 to 1.8 MJ per day.

7. Milk production in well nourished women averages 750 ml/day for the first 4-6 months of lactation.

8. Severe effects on milk production occur when poor intakes in pregnancy are followed by poor intakes in lactation.

9. Essential fatty acids are becoming recognized as important in pregnancy and consumption of fish oil increase pregnancy duration and size of the newborn.

10. Exclusively breastfed infants are at risk of vitamin D deficiency of maternal vitamin D status is low.

11. Premature infants are at risk of vitamin E deficiency.

12. Vitamin K deficiency should be treated in mothers before birth or in infants at birth to avoid hemorrhagic disease.

13. Excess vitamin A is teratogenic.

14. Folate intake of 400 microgram/day is recommended before conception and for the first 3 months of pregnancy to avoid neural tube defects.

15. Vitamin C requirements in lactation are 50% above these for pregnancy.

16. Infants are more vulnerable than their mothers to vitamin B_{12} deficiency.

17. Calcium is important for the fetal skeleton and may reduce gestational hypertension and preeclampsia. Supplements may be needed to achieve recommended intakes.

18. Iron requirements in pregnancy care 30 mg/day, so supplements may be needed. Lactation requires only 15 mg/day.

19. Iron deficiency is common in developing countries and is associated with pre maturity, low birth weight and increased mortality.

20. Women in iodine deficiency areas must have iodine prophylaxis to avoid serious long term mental problems in their children and the whole community.

INFANCY, CHILDHOOD AND ADOLESCENCE

1. Breast milk is a complete and ideal food for the first four months of life. It contains adequate energy, water and all the essential nutrients it promotes development of the infants immunity and it protects the infants from infection and probably from allergic diseases.

2. Suckling is important in the bonding of mother and infants and in the psychological development of the child.

3. Breastfeeding is contraindicated when the mother is taking certain drugs or is resident in the UK and infected with human immunodeficiency virus (HIV).

4. Goat, sheep and cows milk are unsuitable for young infants, as the renal solute load is too high.

5. Formula milks provide a nutritionally adequate substitute for breast milk but cannot provide the immunity and long term health benefits of breast milk bottle feeds risk contamination where hygiene is poor and temperatures and humidity are high.

6. In cases of cow's milk intolerance specialized cow's milk formula containing hydrolyzed proteins is preferable to soy milk.

7. Weaning should not begin before 4-6 months.

8. Ideal weaning foods are more energy dense than milk, 30-40% fat and rich in protein and iron.

9. New foods should be introduced in small quantities at intervals of several days, and fed slowly.

10. Food should be crushed or cut up into small pieces until teeth eruption is complete.

11. Infants need to be fed more than three meals a day.

12. Cow's milk should not become the main drink until the second year. Skimmed milk should not give to children under 5 years.

13. Low iron intake is common in children and has serious consequences. Attention should be given to dietary sources of iron. If red meat is not eaten, sources of ascorbic acid such as orange juice aid iron absorption from plant foods.

14. Early childhood is an important time for children to try new foods and develop normal dietary habits and behavioUr; conflict over meals should be avoided.

15. Children of school age become more independent of parental control of food intake, are susceptible to commercial advertising and peer pressure and are at risk of consuming inappropriate foods of limited nutritional value.

16. These high-fat, high-sugar foods are associated with increasing obesity and poor dental health.

17. Adolescence is the time for the final growth spurt and the attainment of peak bone mass. Sexual maturity brings a major risk of anemia and in girls the risk of unplanned pregnancy. Food habits often change drastically and yet optional nutritional is crucial to feature health.

18. There is serious need for improved nutrition education at school, wide ranging policy measures and research on the impact of childhood nutrition on adult disease.

NUTRITIONAL MANAGEMENT OF DISEASE OF THE GUT

1. Acute diarrhea due to intestinal infection is a major cause of death in children. The salt and water can be treated by oral rehydration solution (ORS), which combines salt and an absorbable carbohydrate. The ideal carbohydrate source has not yet been identified but low osmolarity solutions may be preferable.

2. Oesophageal problems which lead to choking, coughing and difficulty in swallowing, can be helped by sitting up right for feeding, allowing plenty of time to eat foods that do not need lots of chewing, and washing solid foods down with liquid. In more severe cases a percutaneous endoscopically inserted gastrostomy (PEG) may be needed.

3. Gastro-oesophageal reflux disease (GORD), of which heartburn is major symptom, is managed largely by non-dietary means. However, maintaining ideal weight avoidance of alcohol, coffee, chocolate, fatty foods or large meals at night may be helpful.

4. Peptic ulcer, which is one of the major causes of indigestion or dyspepsia, is now known to be due to gastric infection with Helicobacter pylori and is treated with a 1 week course of antibiotics and suppression of gastric secretion. Diet has very little role in therapy but this is an opportunity to provide patients with advice on a prudent diet.

5. Surgery of peptic ulcer has become much less common however gastric surgery can lead to unpleasant symptoms and syndromes. Management lies in preventing hypertonic material entering the upper intestine by avoiding surgery foods and drinks, milk and alcohol. Carbohydrate should be as starch, provided in small, frequent meals. Supplements of Fe and vitamin B_{12} may be needed and vit. D levels and exercise should be maintained to prevent osteomalacia.

6. Malabsorption, which is mostly manifest as steatorrhea, results from many conditions. A normal diet should be taken as far as is possible, although low fat diets and supplements with medium chain triglycerides or individuals minerals and vitamins may be needed for specific cases.

7. Coeliac disease is caused by an abnormal sensitivity to cereal gluten in genetically susceptible individuals. Diet is essential in the management to these patients and requires exclusion of wheat, barley, rye and oat products for life. Rice, maize and soy are safe. Some celiac patients however can tolerate oats. Wheat flour is added to many foods, so patients must learn to observe food labels.

8. Dermatitis herpitiformis also responds to gluten exclusion.

9. Tropical sprue is treated with broad spectrum antibiotics and folic acid. A good diet adequate in energy, protein and micronutrients hastens recovery.

10. Surgical resection of the proximal small bowl usually results in compensatory hypertrophy of the remaining distal bowel. Ileal resection results in bile acid and vitamin B_{12} malabsorption. If of moderate degree it can be managed with the bile acid binding resin cholestyramine and prophylactic vitamin B_{12} injections. Larger resections require a low fat diet and control of gastric acid hyper secretion.

11. Massive small bowel resection (loss of 75%) requires initial parenteral feeding but eternal feeding should begin as soon as possible to prevent mucosal atrophy. As with gastric surgery avoid hyper tonic solutions especially glucose and lactose. Give small, frequent, low fat meals. Dietary supplements of Fe, Zn, essential fatty acids (EFA), Ca, vit. D and injections

of B_{12} will be needed. Oxalate renal stones, D-lactic acidosis and bone disease are recognized complications.

12. Lactose intolerance does not always require dietary intervention because these subjects can tolerate up to 10-15 g lactose per day quite well. A low lactose diet requires reduction of milk intake together with ice cream and some cheeses. Many types of yoghurt have lactose added to them.

13. Crohns disease and ulcerative colitis are the principal inflammatory bowel diseases. In both conditions inflammation is controlled with steroids, 5-amino salycylates or immuno suppressive drugs. Surgery may be needed, particularly in crohns disease.

14. Diet is valuable adjunct to the treatment of crohns disease. Anemia and weight loss are common and specific mineral and vitamin deficiencies occur with malabsorption. Bone disease is well recognized. For actually in patients, eternal feeding with elemental diets is a valuable adjunct to conventional therapy. Exclusion diets have been found to be valuable in some patients in the longer term.

15. Diet is not currently thought to have a major role in the management of ulcerative colitis.

16. Constipation has many causes. The best management strategy is to allow plenty of time to go the toilet. And to increase non-starch polysaccharides (NSP) intake, either through the use of bulk laxatives or dietary modification.

17. Irritable bowel syndrome (ISB) is very common. It probably represents a mixture of pathologies and no single treatment is universally beneficial. In some cases food intolerance may be alleviated by an exclusion diet and where constipation is a problem, increased NSP intake can be valuable. A low gas diet may be helpful.

18. Diverticular disease may be managed with a high fiber diet. Antispasmodics and antibiotics may occasionally be needed to control pain or infection.

19. A high fiber diet to avoid constipation is the best treatment for hemorrhoids and anal fissure.

20. Stoma patients should eat full and varied diet, chew food thoroughly avoid salt and water depletion, and be aware that fish, onions, leeks and garlic cause odour and peas, beans and fizzy drinks produce excessive gas. It is best to eat more in the morning and at midday and less in the evening.

NUTRITION AND THE LIVER

1. Patients with liver disease are often malnourished.

2. Nutritional status should be assessed in all patients initially and should be monitored regularly.

3. Protein and energy requirements are increased in patient with cirrhosis: minimal daily intakes of 30 kcal/kg and 1 g protein/kg are recommended.

4. Patients with chronic liver disease should be encouraged to take 4-7 snack meal throughout the day with a late night snack of complex carbohydrate.

5. Dietary sodium intake should not be restricted to below 60-80 m mol in patients with cirrhosis and as cites long term.

6. Dietary protein should not be restricted in patients with cirrhosis and hepatic encephalopathy.

7. There is much evidence that obesity promotes gallstones, in younger women it is the dominant factor.

8. Fat which is abdominally distributed is the main culprit in men and possibly women too. Obesity may operate through insulin resistance and hyperinsulinamia.

9. Crash slimming diets are a potent cause of gallstones. Otherwise, the case against the specific eating habits are dietary components is inconclusive.

10. There are however some grounds for incriminating extrinsic sugars and rapidly digestible starch, again perhaps operating via hyperinsulinamia.

11. A low intake of dietary fiber 'ought' to be causative because of the clear links between gallstones and the bacterially derived bile salt deoxycholate and between gallstones and slow intestinal transit, however epidemiological evidence is contradictory.

12. Drinking alcohol in moderation is probably protective.

NUTRITION AND DIABETES

1. There are several types of diabetes mellitus, all characterized by an elevation of blood glucose when fasting and following an oral load of glucose.

2. Type 1, insulin dependent diabetes mellitus (IDDM) is a disease of insulin deficiency. It requires insulin treatment; food particularly carbohydrates should be eaten at times which match the insulin delivery.

3. Type 2, Non insulin dependent diabetes mellitus (NIDDM) is a disease of nutrient storage and insulin resistance strongly linked with overweight.

4. Development of NIDDM in middle age can be prevented an early stage

NIDDM is reversible. With weight loss and physical activity. Weight control should be targeted at the whole family.

5. Oral hypoglycemic drugs enhance insulin action stimulate insulin secretion or delay carbohydrate absorption.

6. Sulphonylurea drugs may provoke hypoglycemia and their use requires regular timing of meals and snacks.

7. The goal of dietary advice to diabetics is to maintain health and quality of life and avoid the vascular complications of diabetes.

8. The pattern of meals should be providing food every 2-3 hours. This is best achieved by three regular, well spaced meals, with three healthy snacks in between.

9. Estimates of energy requirements should be based on body weight age and activity levels.

10. Moderate exercise is beneficial to otherwise healthy IDDM patients,.

11. Extra food is needed after exercise in IDDM. Insulin reduction is advised if exercise is planned.

12. In IDDM weight gain increases the need for insulin, weight loss reduces requirement. A strategy of no weight gain with age is earlier to fulfill than weight loss.

13. A diet lower in fat and higher in carbohydrate may be all that is needed to control weight.

14. Dietary advice for diabetics is similar to that for the general population and is particularly relevant to the avoidance of vascular complications.

15. Carbohydrate should provide more than 50% of energy and should come from complex sources containing fiber.

16. Soluble fiber such as gums, gels and pectin's in natural foods are beneficial in modulating glycaemia.

17. Foods of low glycaemic index (GI) such as pulses are valuable in controlling glycaemia and in lipid reduction.

18. Sugars, while helpful in treating hypoglycemia, have no other benefit and should comprise less than 10% of energy and be counted as part of the total carbohydrate intake.

19. Fructose and other nutritive sweeteners have no special advantages for diabetics.

20. Non-nutritive sweeteners are useful in drinks and in meal planning for the overweight.

21. Special diabetics' foods are expensive and have no health benefits over normal foods.

22. Protein intakes should be close to official recommended levels at about 0.8 g/kg body weight per day.

23. In NIDDM, alcohol should be limited to for instance, one or two glasses of wine a day and taken with food. In IDDM alcohol impairs glycaemic control and aggregates hyper triglyceridaemia.

24. Vitamin supplementation is not generally recommended in the treatment of diabetes except in cases of marked deficiency but diets rich in all vitamins; particularly antioxidants in fruit and vegetables, and in minerals, especially magnesium and zinc are encouraged.

25. Salt restriction to less than 6 g/day is recommended for the general population and is particularly relevant in diabetes.

IODINE DEFICIENCY DISORDERS

1. Iodine is an essential element because it is a constituent of the thyroid hormones; thyroxin (T_4) and triiodothyronine (T_3) which are essential for normal growth and development in man and animals.

2. The effects of iodine deficiency on growth and development are known as the iodine deficiency disorders (IDD). Most common is goiters at all ages, but most significant are the effects on brain development during pregnancy and the first two years of infancy which are periods of rapid brain growth.

3. The major effects of iodine deficiency is endemic cretinism, characterized in its fully developed form by mental defect, deaf mutism and spastic diplegia and which can be totally prevented by correction of the iodine deficiency before pregnancy. The extent of brain damage that occurs due to iodine deficiency in a given population varies considerably, but a Meta analysis of studies reveals a mean loss of 13.5 IQ points.

4. Experimental studies in the sheep, rat and marmoset confirm the effect of iodine deficiency on fetal brain development and show a slowing of neuroblast multiplication and differentiation and a consequent effect on neurological development. These effects are due to failure of transfer of sufficient thyroid hormone from the mother during the first half of pregnancy and failure of fetal thyroid hormone secretion in the second half.

5. Iodine deficiency is now recognized as the most common cause of preventable brain damage.

6. The condition occurs particularly with subsistence agriculture in association with a deficiency of iodine in the soil due to glaciation and high rainfall in mountains (Swiss, Alps, and Himalaya) and flooded river valleys. (India, Bangladesh and China).

7. The population at risk is estimated by WHO to be 1.6 billion of whom more than 20 million are estimated to have some degree of preventable brain damage.

8. In many industrialized countries iodized salt has been effective in the correction of iodine deficiency. Difficulties in developing countries with production, distribution and marketing have led to the adoption of a policy of universal salt iodization (USI) which aims to achieve access to iodized salt for at least 80% of the population.

9. Iodized oil, administered initially by injection, but more recently orally, has been used on a large scale to correct iodine deficiency when effective salt iodization is delayed or impracticable.

10. Assessment of iodine nutrition status at the population level relies initially on estimates of goiter rate and measurement of iodine excretion in the urine. Ultrasonography is now being used to provide an objective measure of thyroid size. A population is defined to have a public health problem if more than 5% of children aged 6-12 years are found to have an enlarged thyroid.

11. WHO recommended levels for salt iodine are in the range 20-40 mg per kilo. It is most important that the median urine iodine exceeds 100 microgram per I to prevent iodine deficiency, (and protect the fetal brain) but is not above 200 microgram in order to minimize the occurrence of iodine induced hyperthyroidism. IIH has been observed when inadequate mixing of iodine in salt has led to excessive levels. Only regular monitoring of salt iodine and urine iodine levels will ensure a satisfactory maintaince of normal iodine nutrition.

12. Criteria for monitoring progress towards the elimination of IDD as a public health problem have been established by WHO, UNICEF, and the ICCIDD. The elimination of iodine deficiency disorders by the year 2000 was accepted as a goal in 1990 by the World Summit for children at the United Nations in New York and by the World Health Assembly in Geneva. These commitments have led to high level political support by Heads of State.

13. Since 1990 there has been remarkable progress with the adoption of universal salt iodization programme by many governments. By mid-1996 it was estimated by WHO that approximately 56% (2-4 billion) of the population in 83 developing countries with a significant IDD public health problem and access to iodized salt.

14. A resolution from the 1996 World Health Assembly notes this remarkable progress and urges increased efforts for the sustainability and elimination of IDD by continued monitoring, training and technical support with the International Council for control of iodine deficiency disorders and

UNICEF. A further report on progress is to be made to the 1999 World Health Assembly.

15. The elimination of IDD as a cause of brain damage by the year 2000 would be a greater triumph in a quantitative terms than the previous triumph of the eradication of smallpox. Progress since 1990 suggests that it could be largely achieved, and so conquer an ancient scourge of mankind.

SKIN HAIR AND NAILS

1. Dietary imbalance can disturb the equilibrium of the skin.

2. Skin disease can cause metabolic stress which may alter the demand for nutrients.

3. Kwashiorkor results from severe protein energy malnutrition combined with zinc and multivitamin deficiency and causes severe changes to skin and hair.

4. Vitamin A is essential to healthy skin and has an antikeratinizing role. Its therapeutic role in congenital skin conditions is limited by its toxicity in high doses.

5. Deficiency of any of the B complex vitamins has multiple skin effects.

6. Niacin deficiency is possible in those whose staple diet is maize, and cause pellagra with severe dermatitis.

7. Riboflavin and pyridoxine deficiency cause skin lesions, particularly at the angles of the mouth, in flexures and genital regions.

8. Vitamin C deficiency causes scurry, with multiple effects due to faulty collagen formation, obvious symptoms are spongy, bleeding gums and hemorrhaging around follicles particularly in lower limbs. Response to treatment with ascorbic acid is rapid.

9. Acrodermatitis enteropathica is a rare, severe congenital disease with alopecia, vesicles and exudative dermatitis; it is the result of zinc deficiency despite normal intake. Treatment is with high dose (220 mg) zinc sulphate daily with food.

10. Iron deficiency may result in a range of skin symptoms including spoon nail and itching.

11. Iron overload may cause the skin to be grey and cause hepatic siderosis. This can be treated by venesection.

12. Essential fatty acid deficiency increases transepidermal water loss in animals and causes scaling of the skins in the humans; topical applications of EFA reverse the symptoms.

13. Treatment of eczema with topical or dietary EFAs yields conflicting results.

14. Treatment of psoriasis with fish oil, topically or internally and in combination with other treatments may confer benefit, but results of trials are inconclusive.

15. Urticaria or skin rash caused by a local histamine reaction is commonly linked to consumption of milk, eggs, fish, shellfish, nuts and strawberries with allergy arising in early childhood. The foods that cause the response should be avoided. Evidence that breastfeeding is protective against allergies is equivocal.

16. Dermatitis herpetifirmis may respond to exclusion of gluten from the diet.

NUTRITION AND AGING

1. Chronic disease and poor health is not inevitable in the elderly.

2. Loss of muscle and decline in strength, mobility and agility with age can should be minimized by approximate diet and exercise, in sunshine whenever possible.

3. Folate and vitamin B_{12} are important in prevention of cardiovascular and peripheral vascular disease.

4. Cognitive function is dependent on adequate vitamin supply particularly B_{12}.

5. Loss of lean tissue results in lower energy requirements so the diet needs to be nutrient rich to obtain adequate micronutrients.

6. The diet of the elderly should be rich in unrefined cereals and bread, green, yellow and orange vegetables and fruit, legumes and dairy products and regular small portions of fish, poultry, meat or egg, to provide all these nutrients in natural forms.

7. Vitamin D, B_{12} and calcium may be needed as supplements by those at special risk of deficiency.

NUTRITION AND THE IMMUNE SYSTEM

1. Any component of the immune system may be absent or abnormal the effect may range from trivial to fetal.

2. Dysfunction of the immune system may be genetically determined or acquired from disease drug treatment or radiotherapy.

3. Immune deficiency may result in under activity of the system and susceptibility to infection or abnormal regulation as in allergy or autoimmunity.

4. AIDS is characterized by severe weight loss resulting from anorexia and malabsorption by an infected small intestine. Advice on nutrition with nutrient rich foods and lactose exclusion gives the best practical support. Advantage should be taken of arabolic acids, when infections are under control, to boost feedings.

5. Poor nutrition compromises immune function not only in the malnourished of developing countries, but also in the undernourished of industrialized nations, such as those in hospital with infection, surgery, trauma or cancer, alcoholics, the elderly and food faddists.

6. Protein energy malnutrition (PEM) in children causes atrophy of the lymph organs (thymus, tonsils, spleen, lymponodes, peyers patches), with reduction in T lymphocytes, complement production and bactericidal activity.

7. Essential fatty acids (EFA) deficiency also causes lymphoid organ atrophy and the resulting depressed antibody responses.

8. Deficiency of vitamin A, E, B_6 pantothenic acid and folate all decreases cell and humoral immunocompetence, with an increased susceptibility to infections.

9. Vitamin C enhances bacterial phagocytosis.

10. Deficiency of the minerals Fe, Zn, Ca, Mg, Mn, Cu, Se, Cd, Cr, I depresses immunocompetence as do toxic levels of the heavy metals Pb, Hg, Cd.

11. Iron supplements should be given with caution to malnourished patients with infections as iron stimulates bacterial multiplication.

12. Zinc deficiency in fetal and early neonatal stages delays the development of the immune system.

13. Food intolerance and food sensitivity are terms for all reproducible adverse reactions to specific foods or ingredients which are not psychologically based.

14. The mechanisms may be clear such as lactase deficiency for the milk intolerance, or histamine release following the ingestion of shellfish or may be unknown. Allergies involve immunoglobin (IgE) and mast cells.

15. Food aversion, intolerance and allergy are difficult to distinguish and can only be established by an elimination diet and controlled reintroduction of the target foods. If the foods are disguised, the food aversion response will be negative but intolerance or allergy will be confirmed by positive responses.

16. Infants have an immature immune response system so at 4-6 months new foods should be introduced individually several days apart.

17. In adults, prevalence of food intolerance is probably below 1% excluding lactose intolerance.

18. Lactose intolerance can be managed with a low lactose diet. Fermented milk and lactose reduced milk products may be used to supplement small amount of milk.

19. Doctors and specialists should be sympathetic and supportive to patients with food intolerance to avoid patients resorting to unscientific practitioners.

DIET AND CANCER PREVENTION

1. Obesity is associated with a marked increased risk of endometrial cancer and with a greater risk of postmenopausal breast cancer, and bowel cancer in men. Overweight should therefore be avoided.

2. Low fruit and vegetable consumption is associated with increased risk of cancer at virtually all sites. Consumption probably needs to double, partly to meet existing recommendations to increase non starch polysaccharide (NSP) consumption by 50% from 12-18 gm/day. This will increase stool weight of 25% and reduce large bowel cancer incidence on a population level by 15% vegetables may have added benefits in reducing cancer rates because they contain antioxidants nutrients and flavonoids, sulpher containing compounds and folate all of which can be shown to favourably affect factors thought to be important in reducing risk.

3. High fat diets may enhance colon cancer risk. Recommendations to reduce total fat intake to 33% energy and to increase ω-3 fatty acid consumption in order to reduce the risk of cardiovascular disease (CVD) may be expected also to reduce large bowel cancer rates. There appears to be no effect of fat on breast cancer risks in adult life, but low fat diets generally are low in energy content and assist in the avoidance of obesity.

4. Meat consumption, especially red and processed meat consumption is liked with higher risks of bowel, breast and prostate cancer. Consumption of these should not be therefore increase.

5. Alcohol increases the risk of cancer of the mouth, pharynx, larynx, oesophagus and liver in men and women and possibly cancer of the breast in women and there is a multiplicative effect with smoking. Current guidelines for sensible drinking should not be exceeded, particularly by people who have a family history of the disease.

6. The risks from the sunlight exposure and from smoking are well known but other precise dietary factors that may protect against cancer are not so well established. In humans, the evidence so far relates to food, especially to vegetables rather than supplements of vitamins, minerals or purified extracts. Attempts to alter risks with supplements have so far not been successful and are not recommended for the general population. Mouldy foods should be avoided.

PHYTOPROTECTANTS

1. Phytoprotectants are bioactive compounds found in foods of plant origin which may be benefits to human health. One way of grouping them is an organosulphur compounds, products of the isoprenoid pathway and products of the phenylpropanoid pathway.

2. The organosulpher compounds are found in alliums vegetables, such as onions, garlic, leeks and chives and the brassicas, such as cabbage cauliflower, broccoli, Brussels sprouts and kohlrabi. Allyl sulphides may have an antibacterial effect and may also protect against stomach cancer. The glucosynolates in the brassicas metabolize to indoles and isothyocynates which may have a variety of actions relevant to health.

3. Products of the isoprenoid pathway include terpenes, such as limonene which is thought to protect against cancer and carotenoids many of which are antioxidants.

4. The phenylpropanoid pathway is responsible for the biogenesis of several structurally diverse groups of compounds.

5. Many phenylpropanoids have the potential to act as antioxidants, modulate the activity of cytochrome P450 and enzymes in the arachidonic acid cascade, activate phase II reaction and affect cell signaling. *In vitro*, many have been shown to inhibit bacterial and viral replication.

6. Phenylpropanoids include the flavonoids, the catachins, gingerol, the coumarins, plant sterols and phytoestrogens. These groups of compounds may have a variety of therapeutic properties ranging from antiviral effects to protection against heart disease and various cancers.

7. Phytoestrogens have a protective effect against breast cancer, possibly by lowering free oestrogen levels in plasma. They may also have a role preventing postmenopausal osteoporosis.

8   Coronary heart disease (CHD) rates are much lower in regions where soybean protein is eaten rather than animal protein. Antioxidants polyphenols also protect against CHD.

9. Isoflavones and lignums are marketed as health foods, with few indications of ill effects. However there is concern about the high levels of Isoflavones in infants fed on soy milk formulas.

10. Protease inhibitor and saponins may also be associated with benefits to health.

CONSUMER PROTECTION

1. Consumer protection laws were initially enacted to prevent the fraudulent adulteration of food. The increased use of agrochemicals and food

additives in the twentieth century has led to consumer protection legislation which is largely concerned with food safety. Food labeling is also regulated.

2. In the UK and Food Advisory Committee (FAC) is responsible for considering the compositional standards of certain foods. Food hygiene and minimizing the levels of contaminants. Nutritional and toxicological issues are overseen by the Committee on Medical Aspects of Food Policy (COMA) and the Committee on Toxicity (COT), respectively. For the European Union, the scientific staring committee and eight scientific committees are responsible for supplying high quality advice to form the basis of EV legislation concerning consumer health protection.

3. In the UK local government authorities are responsible for enforcing legislation.

4. Within the European community, 'harmonization' of food regulations was first directed at compositional standards, but it is now aimed at food labeling. The E number system is an advanced and unified method for label information concerning additives.

5. In the US, the Food and Drugs Administration (FDA) has developed a strict regulatory code in which the onus is upon the manufacturer to prove that the proposed additive, ingredient or drug is safe.

6. The international codex Alimentarius was established in 1963 by the Food and Agricultural Organization (FAO) and World Health Organization (WHO) of the United Nations.

7. Provision of accurate and informative food labeling is a major concern of consumer protection. Nutritional labeling is seen as part of public health strategy, but it appears that for this to be effective a programme of public education in nutrition is also needed.

8. The main thrust of consumer protection regarding nutritional supplements has been to ensure that they are not over consumed. Claims for health foods are often difficult to substantiate and consumers need to be aware of the potential hazards associated with some of them.

POLICY AND PRUDENT DIET

1. Many governments have to deal with the dual problems of hunger and communicable disease rurally and chronic disease in urban centers.

2. Government strategy should consider whole populations not just high risk groups.

3. More lives and medical costs can be saved by tackling those large numbers at average risk than the smaller number at high risk.

4. Nutrient goals have been devised by international committees and are based on systematic review of evidence from animal, epidemiological and clinical studies from all over the world.

5. Nutrients goal are the national average intake of a nutrient considered appropriate for ensuring optimal health of a population.

6. Nutrients goals have been chosen pragmatically with lower and upper limits to suit different populations.

7. Where a population's current intake is very different from the ultimate goal, intermediate goals may be set.

8. Nutrient goals must be converted by nutritionist and dietitians to culturally and regionally appropriate dietary guidelines.

9. Health education is not enough to cause dietary change. Change in dietary behaviour is best achieved when government policy in economic, organizational and social measures encourages the choice of a prudent diet.

10. Education and policy on nutrition matters must be directed on a multisectorial basis at all levels of the food chain.

REFERENCES

Allured's Flavours and Fragrance Materials, Allured Publishing Corp. Carol Stream, III. 2003.

Anonymous, 1982. **Prevention of Food Adulteration Act.** Government of India Publication, New Delhi.

Chakraverty A., Mujumdar, A. S., Raghavan, G. S. and Ramaswamy, H. S. 2003. **Handbook of Postharvest Technology. (Cereals, Fruits, Vegetables, Tea and Spices).** Marcel Dekker, Inc. New Delhi.

Chaplin, M. F. and Bucke, C. 1990. **Enzyme Technology.** Cambridge University Press, Cambridge.

Code of Federal Regulation, Title 21-Food and Drugs. **Subchapter B-Food For Human Consumptions.** Parts 100-199. U.S. Government Printing Office, Washington , D.C.2003.

Desrosier, N.W. 1977. **Elements of Food Technology.** AVI Publishing Co., INC., West Port. Connecticut.

D.E. Pszczola, Food Technol .57 (11), 48(2003).

FAO, 1973. **Legumes in Human Nutrition.** Food and Agriculture Organization of the United Nations, Rome.

Fennema, O.R. Ed. 1985. **Food Chemistry.** Marcel Dekker Inc. New York.

Fennema, O.R. Ed. 1976. **Principles of Food Science Part-I. Food Chemistry.** Marcel Dekker Inc. New York.

F.J.Francis , Colorants, **American Association of Cereal Chemists,** St. Paul, Minn.(1998).

Garrow, J.S. and James, W.P.T. 1998. **Human Nutrition and Dietetics.** Churchill Living Stone, Edinburgh.

Garrow, J.S., James, W. P.T., and Ralph, A. 2000. **Human Nutrition and Dietetics.** Churchill Living Stone, Edinburgh.

George, F. S., Mayard, and Amerine, A. 1973. **Introduction to Food Science and Technology.** Academic Press, New York, London.

G.A. Burdock, **Food Technol.** 57(5), 17 (2003).

Hulme, A.C. 1978. **Biochemistry of Fruit and their Products.** Vol. I and II Academic Press London.

Haung, M.T., Toshihiko, O., Ho, C.T. and Rosen, R.T. 1994. **Food Phytochemicals for cancer Prevention.** I. Fruits and Vegetables. ACS Symposium Series 546. American Chemical Soc. Washington. DC.

J.W.Looney , P.G. Grandal , and A. K. Poole, **Food Technol.** 55(4), 60 (2001).

Kader, A.A. Ed. 1992. **Postharvest Technology of Horticulture Crops.** University of California, Publication No.3311

Karel, M. Fennema, O.R. and Lund, D.B. Eds. 1975. **Principles of Food Science** Part-II. Physical principles of Food Preservation. Marcel Dekker Inc. New York.

Kharatyan, S.G. 1978. **Microbes as Food for humans.** Ann. Rev. Microbiol, 32: 301-307.

Lal, G., Siddappa. G.S. and Tandon, G.L. 1986. **Preservation of fruits and vegetables.** Indian Council of Agricultural Research, New Delhi.

Liener, I.E. 1980. **Protein Nutritional Quality of Foods and Feeds,** New Protein Foods. AVI Publication Co., INC. West Port, Connecticut.

Liener, I.E. 1969. **Toxic constituents of plant foodstuffs,** (2nd edn.). Academic Pres, New York.

Liener, I. E. 1966. **World Protein Resources.** Advances in Chemistry, series 57, American Chemical Society Publication, Washington, DC.

Longman, O. 1986. **Basic Food Preparation.** A complete Manual Orient. Longman Ltd. New Delhi.

L.O. Brien Nabors and R.C. Gelardi, in **Alternative Sweeteners,** 2nd. Eds. Marcel Dekker, Inc, New York, 1991, pp. 1-10 .

L.O`Bren Nabors, Food Technol 56(7), 36(2002).

Macrae, R., Robinson, R.K. and Sadler, M.J. Eds. 1993. **Encyclopaedia of Food Science, Food Technology and Nutrition.** Academic Press, London.

Magnus, Pyke. 1982. **Food Science and Technology.** Deh Hua Printing Press Co., Ltd, Hong Kong.

Meyers, R.A. (ed.) 1995. **Molecular Biology and Biotechnology.** A Comprehensive Desk Reference. VCH Publishers, Inc., New York.

M.E. Parish and Co-Workers, **Comprehensive Review in Food Science and Food Safety**, Vol. 2, Suppl. 1, 2003, p. 168.

P.M. Davidson and M.A. Harrison, **Food Technol**. 56 (11), 69: 2002.

Rahman, M.S. (ed.) 2007. **Handbook of Food Preservation.** CRC Press, Boca Raton London, pp. 1068.

R.L. Smith and Co-workers, **Food Technol.** 57(5), 46(2003).

Salunkhe, D.K. and Kadam, S.S. Eds. 1995. **Handbook of Fruit Science and Technology:** Production, Composition, Storage and Processing. Marcel Dekker. Inc. NewYork.

Salunkhe, D.K. and Kadam, S.S. Eds. 1997. **Handbook of vegetable Science and Technology: Production, Composition, Storage and Processing.** Marcel Dekker. Inc. NewYork.

Salunkhe, D.K., Bolin, H.R. and Reddy, N.R. 1991. **Storage, Processing and Nutritional Quality of Fruits and Vegetables.** CRC Press Boca Raton, Florida.

Salunkhe, D.K. and Desai, B.B. 1984. **Postharvest Biotechnology of Fruits** (Vol. I and II). CRC Press Boca Raton. Florida.

Salunkhe, D.K. and Desai, B.B. 1984. **Postharvest Biotechnology of Vegetables** (Vol. I and II). CRC Press Boca Raton. Florida.

Salunkhe, D.K., Bolin, H.R. and Reddy, N.R. 1991. **Storage, Processing and Nutritional Quality of Fruit and Vegetables** 2nd Ed. Vol. II. Processed Fruits and Vegetables. CRC Press. Boca Raton, Florida.

Shakunthala, M.N. and Shadaksharawamy, M. 1987. **Foods: Fats and Principles.** Wiley Eastern Limited, Bombay.

Sidappa, G.S. and Tondon, G.L. 1959. **Preservation of Fruits and Vegetables.** Indian Council of Agricultural Research, New Delhi.

Srivastava, R.P. and Sanjeev Kumar, 2006. **Fruit and Vegetable Preservation. Principles and Practices.** International Book Distributing Co., Lucknow. India.

Swaminathan, M. 1988. **Handbook of Food Science and Experimental Foods.** Bangalore Printing and Publishing Co. Ltd., Karnataka.

Trueman, P. 2007. **Nutritional Biochemistry.** MJP Publisher, 44, Nallathambi Street, Triplicane, Chennai.

Techical Literature, Danisco USA, Inc, New Century, Kas. 2003

Technical Literature, PURAC America Inc., Linconshire, Ill. 2003.

Technical Literature, Dianisco USA, Inc. New Century, KAs, 2003.

Technical Literature, Chr. Hansen, Inc. Milwaukee, Wisc. 2003.

Vijaya Khader, 2001. **Textbook of Food Science and Technology.** ICAR, New Delhi.

Yamaguchi, M. 1983. **World Vegetables: Principles, Production and Nutritive Values.** AVI West Port. CT.